TO CATCH THE SUN

Inspiring stories of communities coming together to harness their own solar energy, and how you can do it too!

Lonny Grafman | Joshua Pearce

Humboldt State University Press

HSU Library
1 Harpst Street
Arcata, California 95521-8299

hsupress@humboldt.edu
www.tocatchthesun.com

Cover design by Ana Emma Mejía
Interior design and layout by Marian Voicu of Layouts.ro
Reviewing and advising provided by many Appropedians

ISBN 13: 978-1-947112-62-9
Library of Congress Control Number: 2021947789
10 9 8 7 6 5 4 3 2 First Print

Dedicated to
people like you
working to make the world
a little bit better every day.

Table of Contents

Foreword

To Catch the Sun is a precious collection of story and knowledge that is both practically accessible, and intellectually rich. Lonny sets the framework for the technical knowledge by sharing his journey in practice, often with humbling lessons learned along the way. This open and very relatable honesty is an opportunity for readers to learn critical lessons that are often only learned via personal relationship and experience. The 'Solar Stories' clearly share the intimate human aspect of this work, and the vital importance of taking the time to profoundly listen, observe and, if you are not a part of the community of solar application, serve the work so that your involvement is no longer needed in the end.

The technical knowledge, especially from Joshua's lab, that follows the stories is laid out to give the reader a remarkable depth of understanding in relatively few pages, again with wonderful insight and experience shared throughout. I have never seen a book designed to share the technical knowledge of practical application dive so beautifully into the science of photovoltaics! Each section explores the components, design, financial and safety considerations while offering consistent troubleshooting advice and examples from the field. *To Catch the Sun* is a unique gift for those seeking autonomous solar power for their families and communities!

Dr. PennElys Droz has worked for over 20 years in service of the re-development of thriving, ecologically, culturally and economically sustainable and resilient Indigenous Nations, primarily through the Indigenous organization Sustainable Nations and most recently, within NDN Collective.

This book offers something to everyone from the "solar curious" to the experienced solar designer. You don't need an engineering degree to understand the book—it is approachable by anyone that has a willingness to learn. The book expertly blends technical, economic, theoretical, practical, and ethical solar power topics in an easy-to-read, engaging, and even entertaining way. The book covers all major components of grid-tied and off-grid systems, including solar panels, charge controllers, batteries, and inverters. It gives the reader the information they need to know to design solar systems for their home, cabin, RV, or elsewhere. It describes common mistakes to avoid, different design approaches, and demystifies many of the sometimes confusing concepts of solar power. Chapter 2 is must-read for anyone who is considering implementing solar power in at-risk, energy-impoverished communities locally or abroad.

Dr. Henry Louie is a Professor in the Department of Electrical and Computer Engineering at Seattle University. He is author of the textbook *Off-Grid Systems in Developing Countries*. Dr. Louie was a Fulbright Scholar to Copperbelt University in Kitwe, Zambia and is co-founder of the non-profit organization KiloWatts for Humanity which implements off-grid solar systems in Sub-Saharan Africa

1. Introduction

We, Dr. Joshua Pearce of Western University in Canada and Lonny Grafman of Humboldt State University, co-authored *To Catch the Sun* to bring our diverse backgrounds (Joshua more academic and Lonny more street), and long history of working, in solar power together for more long-term impact.

1.1. From Joshua

We all want electricity to work for us in various ways to make our lives better – whether to power our laptops, turn on lights, run refrigerators, or make our water safe.

Although the world is a giant mess, so many of our problems – from resource wars to climate destabilization - could be solved if we had a low-cost, plentiful, and green source of energy. Luckily, we have it. Photovoltaic devices turn sunlight into electricity, almost by magic. They have come down so far in cost recently that if you build the system yourself, it is likely to be the least expensive electricity available – no matter where you live. Solar energy is abundant, the fuel is free, it does not pollute, and it is available everywhere that humans call home. You have got to love the sun! People don't flock to the beach for nothing.

Figure 1.1
Love the sun.[1]

1.2. From Lonny

In 2018, I published *To Catch the Rain* about communities coming together to catch their own rain. The book was crowdfunded and well received. In the three years since it was published, hundreds of people have shared their stories of rainwater catchment. Community members in over a dozen countries have built rainwater catchment projects based on the book, projects such as a school in Northern California and an orphanage in Haiti both providing rainwater to their gardens for their students' food and education.

---

1 *https://www.appropedia.org/File:Sunset_hand_heart_shell.png*

I wrote *To Catch the Rain* because I love what I do. I get to travel the world and work in communities collaborating to meet their own needs. Traveling is amazing, but/and I often realize we can have more impact working in our own communities (where we know the language, weather, customs, best locations for supplies, sources of funding, and best food spots). To that end, I do over half of my projects in places where I live long-term. In both cases, I feel limited with how much impact I can have working with my own hands. This book, *To Catch the Sun*, like the last book, is a way for me to vicariously live through the impact you will have when you build your project. Thank you for that.

Figure 1.2
Community members from Las Malvinas, Dominican Republic working with community members and students from La Yuca, DR and Humboldt, California to build offgrid solar. I am the photographer.check out those camera angles and solar components!

For *To Catch the Sun*, Dr. Joshua Pearce and I have partnered to bring small scale solar power into the hands of more people. We hope you can use this book to take more power back into your own community.

1.3. Who this book is for and how to read it

The sun lands on us with incredible power. Yet we often find ourselves without sufficient power for our needs. This book is for anyone looking for inspiration and capability with small-scale solar power in order to meet their needs. We focus on small-scale, but the learnings in this book can be applied to large-scale micro-grids or even larger solar farms. That said, our focus will be mostly on off-grid systems that are 1 kW or smaller. Some specific examples include:

✳ A small home in a financially rich country

✳ A few homes in a financially poor country

✳ School rooms and community spaces

✳ Isolated loads like electric gates, pumps, and telecommunications equipment

✳ A tiny home or van life

✳ Glamping and backpacking equipment

✳ Emergency supply, e.g., powering an oxygen machine during a power outage

✳ Zombie-apocalypse equipment

✳ Laptop and cellphone chargers

✳ Solar entrepreneurship devices

To Catch the Sun emphasizes adaptability and iteration to meet your needs. It is one of the few photovoltaic books to cover very small systems with and without batteries, in a global context, for everyday designers everywhere.

This book can also be utilized in curriculum so that students have context for learning about electricity, power and energy, photovoltaics, spreadsheets, and basic math concepts.

There are many resources and professionals out there to help in building larger systems. This book will mostly focus on the small-scale, distributed, resilient systems that you can build yourself. In addition, this book is meant to be a deep knowledge starting point. Ultimately, you might find a video online that is exactly what you want to build. This book can help you determine what to build, what to avoid, and assess if the video is accurate. Building a deeper and broader understanding will help you leverage the most current research, resources, blogs, YouTube videos, etc., so that you can adapt to your, and your community's, specific needs.

The book is separated into a few main sections: Section 2-Solar Stories covers examples of, and learnings from, photovoltaic projects that Lonny has worked on; Sections 3-Solar concepts and 4-Electricity cover the basic concepts in solar and electricity needed to understand a photovoltaic system; Section 5-Components covers the necessary components to design and build your system; and Sections 6 and 7 walk you through the iterative process of sizing your system.

2. Solar Stories

Lonny worked on his first useful solar systems as a homeless teen hanging out in squats and eventually more intentional communities. To him, distributed resources and self-sufficiency were rebellion. He didn't think that he would ever be hired to do solar and was mostly just trying not to get arrested for doing it. Somehow, he weaved, with luck and perseverance, a circuitous path to a life involving designing renewable energy and other distributed resource systems with community members around the world. The following are a few of those stories in his voice.

2.1. Schatz Energy Research Center, US

At the turn of the century (the year 2000), I landed my first internship doing solar power with the Schatz Energy Research Center (SERC) through a program called University-National Park Energy Partnership Program (UNPEPP).[2] I was elated to be getting paid to design photovoltaic systems for the state and national parks. I poured my heart and soul into that project. The lab was filled with brilliant eccentric designers whom I am still close with today. Our project was to design a photovoltaic replacement for two 16 kW diesel generators for a ranger station on a pristine lagoon. We also took on the task of designing a solar powered public bathroom.

---

[2] You know academics named something when it sounds like un-pep! Dr. David Narum is the one who brought the program to SERC. See this peer reviewed journal article on the project and program: Chamberlin, C. E., Lehman, P. A., Sorensen, A. H., Engel, R. A., & Sorter, A. C. (2004). UNPEPP: Bringing Renewable Energy to Redwood National Park.

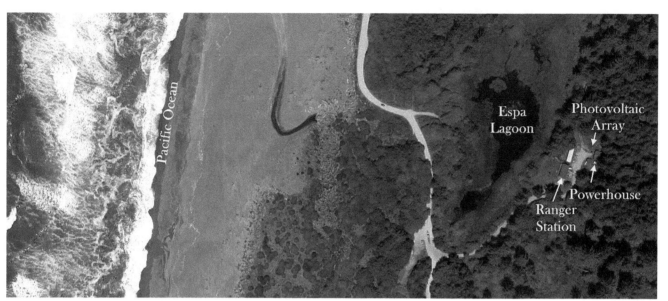

Figure 2.1
Location of the ranger station on the pristine Espa Lagoon in Prairie Creek, California. Image, courtesy of Google Maps, shows new location of the photovoltaic array.

For the ranger station, we first needed to know how much energy they were using and for what. Some aspects were mission critical, e.g., communication equipment that could never afford down time, and other aspects were a little more flexible like a TV and a toaster. We placed cards next to all the devices asking the rangers to write down how many minutes they used each, each time they used it. We also placed energy monitors on the mains coming into the ranger station from the generators.

At the end of the evaluation period, we completed the energy audit and found that the rangers had recorded usage that was much higher (by a factor of two!) than what our energy monitors recorded. We interviewed the rangers and found that they each were overestimating their usage because they wanted to be safe. That was an important lesson in human behavior. Had we not also recorded with large expensive in-line sensors, we would have been designing a much more expensive system to try to meet the rangers' inflated needs.

We finished the ranger station with enough time to really dial in the bathroom. I put all types of creative thought into the design. I wanted it to be not just practical but also an inspirational and educational experience. This was before many people had experienced solar power, and I wanted to share that experience. I put the solar panels high, but front and center so you knew you were entering a solar bathroom. Inside, the wires were in exposed conduit, which was boldly labeled with arrows showing the direction of energy. The charge controller was exposed with its display. The lighting was cutting edge (at the time) LEDs, and all DC to avoid the efficiency loss and costs of inversions. Motion sensors and natural lighting kept the energy costs low. It was beautiful.

At the end of the design internships, we presented to some officials of the State and National Parks. Throughout the presentation, a big man, with a bigger mustache, sat back in his chair with his feet crossed up and his arms crossed in front of himself. I knew the look. At the end of the presentation, we asked for questions. The mustachioed man said, "It won't work." First of all, that is not a question. The old me would have started digging right into him. I was street, but I was learning to fit into academia. I had been fighting his type for years… trying to change the minds of haters and people that did not believe in solar power, climate change, or distributed resources. They would call me a hippie, unrealistic, and a pipe-dreamer. I would call them stronger words. However, here I was, with my first dream internship, trying to fit in and also trying in earnest to figure out how to change people's minds.

I asked him calmly, "Do you care to elaborate?" Alright, in retrospect I am sure my frustration showed, but at least I was trying. He did elaborate. He explained that the exposed system was a target, and that within the first week it would be broken by rocks or even gunshots. That is when I realized that he was right, and I was wrong. I had allowed my optimism and excitement to cloud my vision of the environmental reality. I also did not really learn about the potential users. I never observed, interviewed, or tried to understand any of the local users. I had drunk the renewable energy Kool-Aid and assumed everyone had. This is a lesson that I would unfortunately need to learn many times.

We went back to the drawing board and quickly redesigned the solar bathroom to hide its beautiful secret. At the end of that internship, the National Parks made a brochure of our work spreading the knowledge, and I went back to working outside of renewable energy… for a bit.

2.2. Chiapas, Mexico

My first year teaching a university course was back in 2003, and I landed the perfect solar project! It was exactly the type of project I had been hoping to work on for years. Our task was to design and build a mobile medical station.

The users were rural nurses in Chiapas, Mexico (a state which ranks among the lowest in Mexico on most socio-economic indicators). The nurses would go out to rural villages where the population did not have access to medical facilities. They would travel by truck for hours and then hike for another few. Once with the villagers, they would conduct interviews and tests. Some of the tests they could administer and diagnose themselves, but others required recording data and returning it to the city where doctors could further diagnose and prescribe so the nurses could return to the villages. This was the early days of tele-medicine, but without the telecommunications infrastructure, it was more of a messenger-medicine. It is hard to remember the days before smartphones were ubiquitous, when rural connectivity was still very low and international communication was costly. That was the time we were in.

A non-profit was partnering with the university to create and deliver a solution. The solution needed to power a laptop, be usable by the mobile nurses in extreme conditions, and stay powered with intermittent, and sometimes nonexistent, electrical power sources. It needed to be carried over the long distances and be durable in the mountains and jungles of southern Mexico. It also needed to be able to store all the information for the doctors to evaluate.

What we designed was a laptop in a customized Pelican case. The case could be dropped 15 feet, was waterproof, and was fairly indestructible. It also housed all the components of the system. The laptop could be powered for over 12 hours without access to an electrical power source, which was rare at the time and was achieved by using a separate sealed, gel-cell battery. In addition, the laptop could be powered by the foldout solar panels included inside, or by a 120 V wall plug, a 240 V wall plug, or from a car cigar lighter plug (the round plug that cigar lighters used to fit in, and now are commonly used for USB sockets). In addition, the system could power telecommunications equipment and could trickle charge a car battery in case the mobile nurse vehicle battery went dead. This was all achieved with a bank of plugs and switches that protected the system. The switches worked in combination to help keep the size and cost down. For example, switch 1 up, 2 down, and 3 down meant charge from the solar panel, whereas switch 1 up, 2 up, and 3 down meant charge from a 120 V wall outlet, etc. If something was plugged in incorrectly, diodes prevented explosion. Of course, fuses were included for protection as well. For ease of use, the instructions were provided in English and Spanish.

The entire system weight with the box was under 35 pounds based upon the NIOSH recommendations at the time.[3] The box was also adapted with a strap that could be carried across the head for a balanced ergonomic experience. We were quite pleased with it. The funders were even more pleased. The idea was to recreate these for use with more mobile medical stations. And…it was a complete failure.

I know why it failed. I know what I did wrong. I know the two classic structural errors that caused this. Many brilliant people, and many more Instagram motivational influencers, will tell you that "Failure is just a learning experience." I am telling you that that is B.S. Failure has real consequences. You can also choose to learn from it, and I hope you do… but it is not "just" a learning experience. Real projects come with real impacts to real people. In the case of international development, these consequences often affect those with the least financial resources. This mobile medical project took up real time and resources that could have been put to better use.

---

3   The National Institute for Occupational Safety and Health (NIOSH) https://www.cdc.gov/niosh/topics/safepatient/default.html

The two main problems were ones that have recurred throughout my life as a designer. The first is a classic systemic problem in non-profits. The second is a classic oversight in engineering design.

The first problem stemmed from the common reality for nonprofits that they need to please a different group of people besides the end users. For-profits don't have this same disconnect. For-profits know if their product is successful based on the end users paying for it or not. Whereas in a nonprofit, the money is usually coming from donors who may have great intentions but almost certainly have a different vision than the end users. Donors may love nifty projects (like they did in our case), but end users may not. I have found that wisdom trumps cleverness when designing outside of your own demographic…which leads into the second problem.

The second problem is painfully obvious. We never interviewed, observed, or directly interacted with the nurses. I can blame this on a variety of conditions, e.g., the lack of connectivity back in the day (remember this was the year Skype came out, but very few people were using it). Human-centered design bases many of its tenants on this situation. You cannot design without first building empathy with and understanding of the users. I will most likely never stop being embarrassed by my oversight. Had I interviewed and observed, I would have found that no one wanted a complicated array of wires and switches, nor a large and heavy indestructible box…nor the multiple ways to charge it. In fact, a simple laptop with an appropriately sized solar panel and a charge controller would have done the job. Now all of that is mostly moot, as modern smartphones do the job well.

2.3. La Yuca, Dominican Republic

In 2011, we worked with Colectivo Revark[4] and the Universidad Iberoamericana (UNIBE) under the care of the Director of Architecture, Elmer Gonzalez. Outside of the University, Colectivo Revark was our community liaison and was critical in all the community engagements—yet our community engagement in La Yuca almost did not happen.[5]

La Yuca is one of the most financially poor urban barrios in the center of Santo Domingo, Dominican Republic. La Yuca contains one school, whose schoolyard also serves as the hub of activities for children. You can touch the walls of houses on both sides of the narrow streets as you make your way through the serpentine labyrinth of the neighborhood. If a moped is coming through, you need to press yourself out of the way. The neighborhood bustles with constant activity and noise, including the comforting endemic rhythms of bachata and dominoes. Overhead, masses of makeshift electricity grid extend into the crevices of every home. These floating balls of electrical spaghetti are worked on by many people of La Yuca as repurposed wires burn out in spectacular light shows (Figure 2.2) and need to be replaced.[6]

---

4   An architectural non-profit lead by Abel Castillo Reynoso, Wilfredo Mena Veras, and Joel Mercedes Sánchez.

5   *To Catch the Rain*, Lonny Grafman, ISBN: 978-1947112049

6   Funny story: When we arrived with US students the first time, some of the students thought they were setting off fireworks to welcome us. It was just the electrical lines on fire. Picture included from a few minutes after our arrival.

Figure 2.2
Electrical lines on fire in La Yuca.

The story we first heard from a beautiful ninety-year-old abuela is that La Yuca was originally a temporary home settled by the workers who built much of the surrounding parts of the city, and then the workers didn't leave. There have been many attempts to push La Yuca and its inhabitants out of the city, but La Yuca has prevailed. Colectivo Revark and UNIBE helped set up our first community meeting with the junta de vecinos (city council) of La Yuca. During that first community meeting, the reception was low energy and the junta de vecinos seemed mostly uninterested in working together. It was only after the pastor understood what we were proposing and restated it with eloquence that the engagement happened. He reiterated that we were not there as a charity organization. We were not there with a "solution." We were not there to drop something off and take pictures. We were there to work and learn together. We were there to seek solutions together. And we were there so that we could all gain knowledge, capacity, and build a better future together.

After that initial meeting, we decided to have a meeting open to the entire community where we could identify our top needs and resources. Some top needs included clean water (some people were spending over 40% of their income on water), more school space (there are more students than can fit in the school), electricity (11% of Santo Domingo re-appropriates their electricity), and jobs (incomes are often just a couple of dollars a day).

The open community meeting was loud and fruitful, especially due to the deep support of the town mayor, Osvaldo de Aza Carpio.[7] We brainstormed dozens of available resources and top needs. Then, we prioritized the top needs into just a few and broke off into small groups to brainstorm solutions to those top needs.

Together we decided to create a wind and solar renewable energy system, a schoolroom from plastic bottles (a style called *ecoladrillo*), and a rainwater harvesting system on top of the new schoolroom we were building together. The school was currently ordering two trucks' worth of water per month, which was expensive and was only enough water to clean the school and to flush the toilet manually at night. The school had no water for hand washing in the bathroom nor anywhere else (a major health indicator).

Together we built a 2000-liter rainwater harvesting system that also included additional storage in an existing cistern, reducing the need for purchasing two trucks' deliveries of water per month to just two per year.

We also designed and built a system with a school in La Yuca to provide the energy for their second story. The second story of the school is used for the grade school classes during the day and for adult literacy classes and church groups at night. The rooms are hot, and the power goes out often. When the power goes out, the classes are over. Often classes will run for only a few minutes before needing to cancel due to the rolling power outages.

We worked over multiple summers to design a renewable energy system for the 2nd story. At first, we were trying hybrid solar and wind. The wind worked, even in the

---

7   Osvaldo was not only the consummate community leader in La Yuca - in the years following this first engagement, he has been instrumental in meetings in other communities and has continued to be a sustainability practitioner and advocate of building technologies in his community and the communities of others.

urban setting with low speed, turbulent wind (which is not ideal for wind power). We managed to build the turbine with local and waste materials such as bike parts and the aluminum sheets used for newspaper printing (Figure 2.3). Unfortunately, wind power is proportional to the cube of wind speed (e.g., a doubling in windspeed can be an eight-fold increase in power), and the hurricanes proved to be a major yearly threat. Each year, the system did not survive. Eventually, between the obstacles to the wind power and the rapidly decreasing prices and increasing accessibility of solar panels, we designed the system to be solar only.

Figure 2.3
A hybrid wind (vertical newspaper looking wings) and solar (small square horizontal panel on top) for a school room in La Yuca, Santo Domingo, Dominican Republic. Appropedia.org/La_Yuca_small_scale_renewable_energy_2012

We worked with local leaders and electricians, Osvaldo and Bernardo, to design a system that provides a minimum of two hours of full power for the second story and is turned on only when the power goes out. The system provides the energy for the much-needed lights and fans to keep the night classes running. When it came time to design the system, the students, based on their learning from G2G (Section 2.4), hid the panel to protect it from theft. During the co-design session, the community redesigned this in a spectacular way. They moved the panel to be incredibly visible and hard to take down through the use of metal and barbed wire. The community designers' philosophy was, place it where everyone can see—that way if anyone tries to steal it, they will be spotted (Figure 2.4). This type of community insight comes from being in the community. What was true for the animal shelter in Dominican Republic (Section 2.4) was not true for this school in the same city. This also taught me lessons about different approaches to security in different communities and settings.

Figure 2.4
Two 140 W solar panels in plain view powering a school in La Yuca, Santo Domingo, Dominican Republic with students Jackson Ingram and Javier Durán pictured in the center. Appropedia.org/La_Yuca_renewable_energy_2014

2.4.   G2G Animal Shelter

Two years after starting projects in the Dominican Republic, we started working with Ghetto2Garden in 2013 to build an energy independent animal shelter. Ghetto2Garden (G2G) is an animal shelter in Santo Domingo, Dominican Republic. As an animal lover, I was excited to be asked by Tomás De Santis of G2G to design a photovoltaic system. A student team of US and Dominican students started on their designs and prototyping. Their results were lovely, creative, educational, and resourceful. Towers of Intermediate Bulk Containers (IBCs) housed the electronics and boldly displayed the solar panels up high (Figure 2.5). The IBC towers were located in the center of the animal shelter so that the equipment could be seen and hopefully engender some inspiration for renewable energy, but also far from the edges to prevent theft. The approximately 100 dogs, many of them unwanted pit bulls, were sectioned off into various packs, and each pack had a tower of IBCs to provide enough light to illuminate each pack.

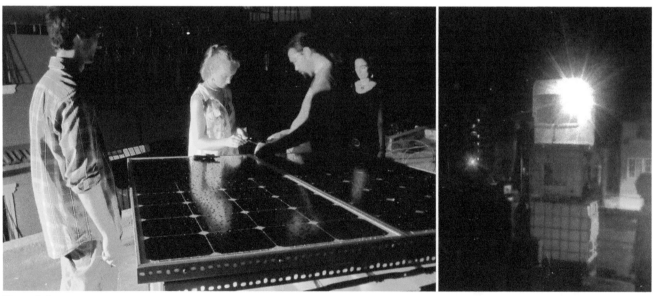

Figure 2.5
Close up of a single IBC with the components hidden inside and a solar panel on top (left). Tower of IBCs with solar light at the top (right).

During the next meeting, we shared with the client the prototypes made of modeling clay. We represented his pack structure as he had described it during our previous interviews. Immediately, he saw the flaw. Our students deflated as he explained that the panels and equipment would be stolen. We countered that the no one could get to the panels without getting through many, many dogs. He shrugged off this argument, with one haunting phrase: "The thieves will just poison all of the dogs." My naivety had struck again. With our newfound, depressing knowledge the students went back to redesigning. Eventually they kept everything except the prominent display of the equipment, which was now hidden flat on top of three levels of IBCs, locked with chain, and wrapped in barbed wire. I had flashbacks to my similar design problem in my internship with SERC (Section 2.1), but this time hiding the panels was much easier because we were closer to the equator where having the panels flat would incur less of a power loss than flat panels would at a higher latitude. More importantly, this time we figured out the error in our design right away through prototyping instead of waiting until the end of the design project to find the error. These specific lessons have countered my naivety periodically: in financially rich environments, vandalism of solar panels can be quite common, and in financially poor environments, theft is often a serious threat.

The final system did provide light for the animals and help with the shelter.

In 2014, one year after the initial G2G solar system, the shelter realized its dream to move outside of the city to the countryside, which came with many new opportunities and issues. The newest issue was that the power needs were higher, e.g., the need for a vaccine refrigerator, and the grid electricity was very intermittent. The power company ran on daily rolling blackouts. This rural community received three hours of electrical energy per day, and no one knew which three hours it would be. Running a shelter, on a hill, now with over 120 dogs and 30 cats, volunteers, and an employee is rendered even more difficult when you don't know which hours you will have power.

Together, now with a history of designing together, we developed a new system. A local electrician and van driver from La Yuca (the community we worked in for the school design in Section 2.3) helped us with equipment, transportation, mangos, and construction. Theft was still an issue and dealt with easily as the site had storage containers that housed the office and storage. The panels would be flat on the roof of these large containers.

The design went quickly based upon our previous learnings. And, the team decided to up the ante by trying to build the system to US code standards, including fuses, ground fault protection, a grounding rod, and a lightning arrestor (Figure 2.6).

Figure 2.6
Site photos (left) mostly included for the cute dogs. Entire system (right) for Ghetto2Garden laid out on the ground for a photo and so we could put the system together for temporary energy to power the tools to mount the system.

After all the design iterations and finding the appropriate equipment, we arrived on site hours outside of the city. There are no telecommunications (cell or internet) at the location. When we arrived, Tomás was apologetic and explained that the power hadn't been on in days and that all of their equipment (e.g., drills) were uncharged. This was one of my favorite solar baller moments. We first built the entire system on the ground and used that to power the equipment to integrate the system into the buildings. Yay for distributed energy systems!

2.5. Baraut, India

Summer of 2018 found me with seven US students (from Humboldt State University in California) and eight Indian students (from Lady Irwin College in New Delhi) living in pairs with families in Baraut (a small city) in Uttar Pradesh (Northern India). I was the co-director of the program with Dr. Meenal Rana, who was the visionary creator of the program in India bringing together amazing partnerships between universities, organizations, and communities. Our host was ESRO (Environment & Social Research Organization), a local NGO (non-governmental organization) based in Baraut, which coordinated our work in two nearby villages: Tavelagarhi and Daula.

Our reception was as warm as the weather. Our second day in Baraut, ESRO held an award ceremony and reception in our honor (Figure 2.7). For days, I had been practicing the opening to my speech in Hindi: "Namaskar! Aapke pyar bhare swagat k liye dhanyawad. Mein aapka abhari hoon. Mera naam Lonny Grafman hai aur mein Humboldt State University mein Engineering ka professor hoon."

Which translates to, in English: "Hello (with respect). I am grateful for your warm welcome. My name is Lonny Grafman and I am a professor of Engineering at Humboldt State University."

It took me a bit due to the sounds that are not common in any of the languages I typically try to speak. The audience applauded, which was quite sweet of them. After that, Meenal Rana started translating for me. I started in English with, "I have been practicing that for a while, and it was quite hard. Except for two words: 'Namaskar,' because that is a common word in California, and 'priya,' because that describes our reception here." I thought that "priya" meant warmth. It did not.in fact it is a relatively common name. Meenal looked at me confused, and asked, "What?" I said, "warmth." She corrected this for everyone. So, during my first English speech in this community, I said that there were only two Hindi words that were easy for me...ones that I thought I had just said...and I was wrong about 50% of them! This became foreshadowing of a few events through the summer. This is also one of my favorite parts of working in new locations. When I am very familiar with a language and an environment, it is easy for me to start thinking I am correct in my language and assessments. I must actively remind myself that I am certainly missing things and need to keep my senses open to conflicting information. I need no such reminding in a new place because the environment reminds me daily!

Figure 2.7
First day honoring ceremony with participants in the summer program.

Northern India in the summer is hot. Our first week there was above 100 °F and very humid. Students were getting used to squatting toilets, constant sweating, new foods, new living arrangements, and most difficultly, new mores. Our group was predominantly women. We had one male student and one male teacher (me). Our students were from some of the most progressive and liberal places in their respective countries. The city of Baraut, and even more so the villages in Uttar Pradesh, are more traditional. Traditional gender dynamics are still very strong and upheld in these locations. We all had a lot to learn, and all our students needed to adapt to the environment.

Luckily, as with all design projects, the very first step is empathy building. Our students' job as travelers, guests, and designers was to observe and understand. Before we arrived, an independent organization (The Center for Environment Communication) conducted interviews with local community members. I was quite impressed with not only their dedication (they provided over 44 hours of interviews) but also their flexibility and understanding. At first, the interviewers suggested to ask the typical leading questions of most NGOs, questions like:

* "On a scale of 1 to 5, how important is energy?"
* "On a scale of 1 to 5, how much of a problem is trash?"

I like these questions; they produce excellent graphs and help with grants and assessment, but they aren't fantastic for empathy building nor design. Instead, we worked together to create design-focused questions, such as:

* "Describe a day you had this week."
* "Describe a time that you had a problem and were able to solve it."
* "Describe a time that you had a problem and were not able to solve it."

These types of open-ended questions are hard to quantify. They are not science. They are empathy building and, in my opinion, much more telling of what your audience wants. Through these questions, we found that the top priorities were water, waste, and energy. An example user-story highlighting the need for energy came from a village girl who described her problem, which translated and paraphrased went like this, "We do not have regular electricity supply. When the power cuts, I am sometimes unable to complete my homework. For this I get punishment in class and not allowed to sit in the class." This type of story helps us to understand the problem and starts to circumnavigate a solution (e.g., very small solar lighting for homework). So, I brought my favorite solar tools, especially some nice multimeters. I try not to bring supplies, as sourcing them more locally brings more long-term success to the project… but I couldn't resist also bringing a few small, efficient panels and my favorite robust, waterproof USB chargers (Figure 2.8.).

Figure 2.8
Small robust solar USB chargers I designed for protests in North Dakota but brought as educational and hopefully inspirational tools in India.

Once on the ground, we conducted our own interviews, conversations, and observations, and most importantly created teams that included a diverse mix of students and community members. After the first two weeks, we abandoned solar as the water and waste issues were more pressing. We did one small community teach-in on solar, and then I put away my solar tools for the summer…that is, until RK Singh, a community member who owns an electronics store in Baraut, suggested we help build his system.

He was going to build the third system in Baraut, this time on his own home. It was going to be a 576 W system with one very unique aspect. The system would have a metal grill above all the panels (Figure 2.9). This, of course, would reduce the efficiency, and therefore power output, of the array… so why have it? RK informed me that it is to prevent the panels from breaking due to monkeys throwing rocks at the panels. I, of course, accepted his reasoning as he has had years of experience with solar power and with monkeys. I was hoping to eventually witness this behavior myself, and often wondered why the monkeys throw rocks at the panels. I thought that maybe it was due to the bright reflection, like how some birds attack windows.

Figure 2.9
Team of installers with the final eight panels and metal protection grate on the roof of RK Singh in Baraut, Uttar Pradesh, India.

Together we assembled the system. Of course, the day was hot, and the roof was hotter. By the end of the install, we made a poor decision and assumed that all the seemingly matching government approved panels we were installing were indeed identical. They were not. It turned out that the positive and negative terminals were switched on some of the panels. Luckily, we did not break anything, and we tested before finalizing the install. With the multimeter, we were able to determine the correct polarity of each of the panels and rewire them for a successful install (Figure 2.10).

Figure 2.10
RK Singh with four of the eight seemingly identical panels, but actually some of the junction boxes have their polarity switched.which we found out almost too late.

During the install, I was teaching the student installers about the efficiency loss from the grill. At the end, RK called me over and asked in the politest English if I thought that monkeys threw rocks at the solar panels. I did. I thought he had taught me that. Through language differences, I had misunderstood. It was in fact to protect against the rocks that people throw at monkeys to keep the monkeys from coming into their home and stealing property. Often, the monkeys are up high when they are plotting their pilfering. Also, the panels are up high on the roof hidden from view, but right in the path of the rock when it misses the monkey… which it almost always does. Now the world made a little more sense. This grill is an interesting story in environmental differences. It is also a telling example of the tradeoffs between criteria such as cost, durability, and power output. Is the approximately 10% reduction in power worth the many extra years of durability? Almost certainly, and the math can tell you for sure.

2.6. Parras de la Fuente, Mexico

Parras de la Fuente is an oasis in the desert of Coahuila in Northern Mexico. As its name (grapevines of the fountain) suggests, Parras is replete with ample spring water and the first winery in the Americas. Being in the desert, Parras also averages over five hours of peak sun daily (with over seven hours in the sunny months and still over four in the rainy months),[8] so it is a great fit for solar.

This political and environmental context was explored through my co-director, Dr. Francisco de la Cabada from Humboldt State University (HSU), our community liaisons from Universidad Tecnológica de Coahuila (UTC), and especially through the work of Carlos Alejandro Ramírez Rincón (director), Simón Leija (environmental science instructor), and Aida Ibarra (business instructor).[9]

In the summer of 2006, we (the Practivistas program: including local community members, students, and faculty of UTC and of HSU) worked with Centro De Salud (the local hospital) in Parras. During our needs and resources assessment, the top needs of the hospital were, on the surface, all projects that we did not have the capacity to help

---

8   This high value of solar irradiation is great for photovoltaic production and covered in later sections.

9   Grafman, L. (2018). *To Catch the Rain*. Humboldt State Press ISBN 978-1947112049.

with. This list of needs included more doctors, more nurses, more money, and more vaccines. Through iterating on these needs, we learned that the need for more vaccines was exacerbated by common power outages that would render the available vaccines useless. and the community would need to wait until the government provided more vaccines. This vaccine expiration due to power outages was a wonderful opportunity for our collaboration in this sunny desert. Even though solar prices were 10x higher in 2006 than in 2021, we were confident that we could co-design an affordable system that met their needs as long as we could leverage some donation.[10]

We started working simultaneously on designing a system and acquiring donations. This system cost needed to be under $500USD, which meant we needed extensive donations. With letters from the hospital and both universities, we obtained donations of panels, a charge controller, shipping, and a very efficient refrigerator for the vaccines from Sunfrost.[11]

Everything was going great until the solar panels got stuck at customs. Customs wanted to charge more to import the panels than the panels were worth. We had the assistance of the local government but still couldn't get the panels past customs without paying. With time running out for my students to be involved with the construction (which would have been fine for the hospital since Antonio, their technician, was part of the design process, but would not have been great for our academic goals), we finally paid customs to permit in the donated panels.

When the panels finally arrived, the program was officially ending and I was still hoping to get the system built with everyone still there. With excitement and apprehension, we opened the package. Quickly searching it for any signs of damage from shipping (and sitting in customs for so long), we turned the panel over only to be met with a ridiculous sign of irony. Instead of letting out a sigh of relief, we double over in laughter. On the back of the panel, a large sticker proclaimed, "Hecho en Mexcio" – Made in Mexico.

---

10   That said, I also started working on a solar absorption refrigerator that used ammonia instead of electrical compression. This process went well until I, in one of my most embarrassing moments, evacuated the UTC campus after pouring ammonia into a solar thermal collector that had become significantly hotter in a short time than I had anticipated. That ammonia vaporized much quicker in the desert sun of Northern Mexico then the coastal fog of Northern California where I had first practiced the pouring. Everyone was okay, thanks to some safety procedures.

11   Sunfrost (http://www.sunfrost.com/) refrigerators were, at the time, the world's most efficient refrigerator and run by the brilliant inventor Larry Schlussler.

We quickly put the system together, except for the refrigerator that was still in transit. It finally arrived after all the students left except one, Jeff Kinzer, who finished the system with Antonio.

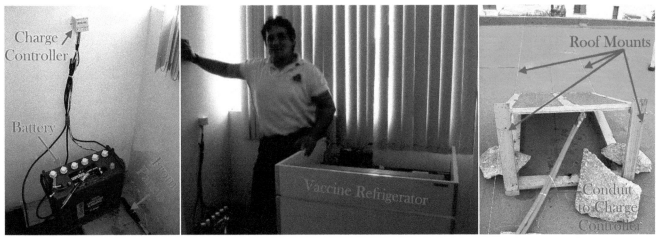

Figure 2.11
Photovoltaic vaccine refrigerator at the Centro de Salud in Parras, Mexico with charge controller and battery (left), solar panels and DIY racking (right), and technician Antonio with the refrigerator (center).

The system worked great and kept the vaccines cold. It was inspiring to see how we were able to brainstorm and work together to meet needs. The students that were part of that project have now gone on to install over 100,000 kW of photovoltaics around the world, such as Aleiha Haley Wayman who is currently helping Hawai'i reach their mandate to be 100% renewable by 2045.

All that said, we could have done better (which is always true). Years after our installation, the batteries were dying, and the hospital technician created an even better system using the "vaccine refrigerator" to make ice that would then keep other vaccine refrigerators cold during power outages. This new system saves significant money in battery storage by using the thermal mass of the ice for storage instead of the electrochemistry of batteries. With all of the donations in our system, the battery was over 50% of the cost…which means we could have built two systems making ice, instead of the one with a battery.

2.7.   Las Malvinas, Dominican Republic

Las Malvinas II, in Dominican Republic, is an ad-hoc community built on the side of the river Isabel next to industry. Many industries set up on rivers, and many ad-hoc communities start with workers moving near those industries. My first visit in 2010 was to a community with strong community ties and governance and severely lacking resources. Since that first visit, all the obvious metrics of human health have improved, especially regarding access to water, education, housing, and employment. It has been a joy to witness the growth and transformation through the work of an engaged *junta de vecinos* run by Ana Sobeida Familia (and Heriberto Eddie Trinidad before her) and so many inspiring community members and organizations.

In the first few years of working together, the community needs were mostly focused on education, health, buildings, and water. Together we built an upcycled schoolroom, rainwater catchment systems, and a community shade structure and invented new building materials like Hullkrete blocks (made from waste rice husks, rice husk ash, clay, sand, and cement). In 2015, energy rose higher on the list of community needs. Based on our learnings from the past and community brainstorming sessions, we decided to run a series of solar workshops with local electricians.

Anyone who self-identified as an electrician qualified and was welcome to take part. Bernardo Martinez turned his house into a classroom and lab. These workshops consisted of some lectures on math and photovoltaic components, and lots of taking small broken solar products and using them to learn about the solar components. In Figure 2.12 is Bernardo's son, Junior Bernardo Martinez,[12] already a great local electrician at age 15, fixing a solar lantern.

---

12   At that time, Junior had already made a Tesla coil from waste parts from the dump. Now, in 2021, Junior is about to graduate university as an Electrical Engineer!

Figure 2.12
Broken solar components being fixed by Junior Bernardo Martinez during a solar workshop in Las Malvinas.

After the electricians learned solar, the team sought to build a solar public pharmacy. The electricians and students designed and built an entire photovoltaic system and used that to teach the rest of the community about solar. Figure 2.13 shows Bernardo teaching about the new, larger 250 W photovoltaic system (left) and students presenting on all the projects (right).

Figure 2.13
Bernardo teaching photovoltaics in the complete 250 W system (left) and students presenting all the community projects in front of the new, planned, public pharmacy building (right).[13]

I love the photos on the left of these last two figures as you can see my red hat getting further back from the action. My job is to become irrelevant, so the further back I am, the more confident I am in the success of the project. The lessons brought to this project were basing it on community needs and engagement, starting small, building upon successes, iterating live with local stakeholders engaged at every step, and supporting local leaders as they ran the larger project and education. People from this and other communities have learned from this project and gone on to build more.

---

13   The government never came through on their promise of a public pharmacy. Now this building and system is being rented by a community bolstering local soap maker - https://www.esperanza-soaps.com/. I am very happy to hear that, and I wish we had done better at ensuring the government's commitment to stocking the public pharmacy.

3. Solar concepts

3.1. Solar power

The sun has used less than half of its hydrogen fuel and should be with us for about 5 billion years.[14] Yet the sun is incredibly, almost unfathomably, powerful. Every single hour, the amount of energy from the sun that strikes the Earth is more than all humans consume in a year.[15] Sunlight has by far the highest theoretical potential of the renewable energy sources on Earth, and there is clearly far more energy coming from the sun than we need. The **solar constant,** which is the solar flux that strikes the Earth, is 1.36 kW/m^2.[16] These numbers are hard to fathom, but you can get a glimpse of the sun's power most easily from images that NASA has taken, like in Figure 3.1.

---

14 Will the Sun Ever Burn Out? https://www.space.com/14732-sun-burns-star-death.html

15 This incredible fact should get you psyched about solar power https://www.businessinsider.com/this-is-the-potential-of-solar-power-2015-9

16 Kopp, G.; Lean, J. L. (2011). "A new, lower value of total solar irradiance: Evidence and climate significance". *Geophysical Research Letters. 38* (1). doi:10.1029/2010GL045777.

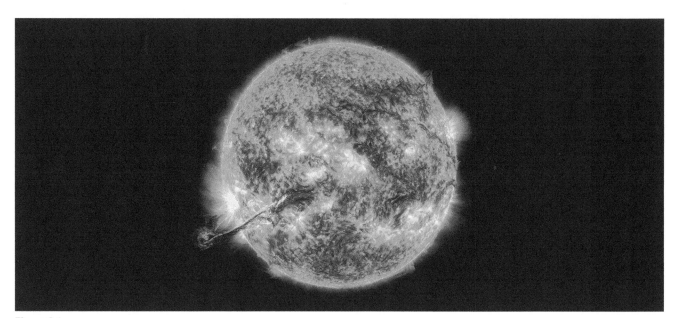

Figure 3.1
The sun – compliments of NASA.

The time-and-space-averaged solar flux striking the outer atmosphere is 342 W/m². Using this value is good for astronauts powering the space station, but about a third of it is scattered and absorbed before us ground-based Earthlings can get to it. For example, ozone (O_3), carbon dioxide (CO_2), and water (H_2O) have very high light absorption properties. This leaves us with about 175 W/m². This is a lot of power – but we still need to capture it. When we do collect the sun's energy with solar cells, this is called **solar power**. Solar power is measured in watts (W). For perspective, the average mobile phone uses approximately 3 to 30 watts when charging, while a charger left plugged in without a phone will consume 0.1 to 0.5 watts.[17] To get a better understanding of the power of the sun, we will discuss how to calculate the power and energy from solar cells and systems in detail in the sizing section.

The sun's power comes in many types, as it is constantly radiating a wide range of colors (or light wavelengths) that go from the UV (or ultraviolet, which are largely cut out

---

17   Electricity usage of a Cell Phone Charger http://energyusecalculator.com/electricity_cellphone.htm

before it gets to the surface – thanks, ozone layer!) to the infrared as shown in Figure 3.2. Sunlight is made up of particles of light called **photons**. When you look at the sun, you see the mix of many photons with different colors. Each individual photon comes in only one color (or wavelength) at a time, and these colors have different energies associated with them. For example, on a clear day at noon, you should apply sunscreen to block the high energy UV photons that are responsible for skin cancer.

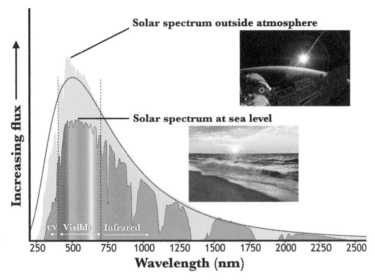

Figure 3.2
The solar spectrum in space and at sea level.[18]

Figure 3.2 shows the solar spectrum – or the flux of photons that come down to the Earth vs. the wavelength (or color) of the photons. The orange line represents the solar spectrum outside the atmosphere. The spectrum reaches a peak in the visible as expected from the physics of black body radiation at 5800 K (this is the surface temperature of the sun, and all objects give off radiation based on their temperature). The missing regions of flux (black line) at the Earth's surface are due to absorption from water vapor and carbon dioxide. As you can clearly see, the solar energy is not just a single wavelength

18 Image from http://www.appropedia.org/Solar_Photovoltaic_Open_Lectures
https://www.flickr.com/photos/gsfc/12867973205 CCBY
https://pxhere.com/en/photo/1026597 CC0

(energy, frequency, or color of light). About 48% of the light is visible to the human eye, 45.6% is infrared, and 6.4% is UV. This makes it a bit tricky to capture all the sunlight with solar cells, as we will see in the next section.

3.2. Photovoltaics

Photovoltaic (PV) systems convert light energy directly into electricity. PV are commonly known as "solar cells." The simplest PV systems power small calculators and children's toys. You can build only slightly more complicated PV systems to charge your mobile phone (Figure 3.3) or power an electric fence (Figure 3.4).

Figure 3.3
Solio hybrid charger powering mobile phone. Image compliments Alan Levine, public domain.

Figure 3.4
*Solar powered electric fence in Oittila village, Jyväskylä,
Finland. https://commons.wikimedia.org/wiki/
File:Solar_powered_electric_fence.jpg CC-By-SA*

Slightly more expensive PV systems can provide power for computers and other electrical devices in a library, school, house, or apartment building (Figure 3.5). These PV systems can be mounted on rooftops or walls as well as in the ground on racks or poles. This allows PV to operate to produce electricity with no impact on the human population in the area.

Figure 3.5

Solar panels on the roof of one of the buildings of the College of Engineering, KNUST, Kumasi, Ghana. https:// commons.wikimedia.org/wiki/File:Solar_panels,_KNUST.jpg CC-By-SA

Finally, larger systems as shown in Figure 3.6 called **solar farms** already provide a portion of the electricity for the world's major grids, and that portion is growing rapidly, as PV is the fastest growing source of power in the world.[19]

---

19   Solar energy is fastest growing source of power. USA Today 2017. https://www.usatoday.com/story/money/2017/10/04/solar-energy-fastest-growing-source-power/730594001/

Figure 3.6
Large solar PV farm - utility scale behind a fence in the US.

Photovoltaic technology holds a number of unique advantages over conventional electric power-generating technologies.

PV is a true **clean and green technology with numerous environmental and health benefits.**[20] PV panels do not require the use of fossil fuels such as coal, oil, or natural gas in the energy production process. In addition, **PV modules produce far more energy in their lifetime than it takes to produce them.**[21] PV can thus be viewed as an energy breeding technology that should be grown as rapidly as

20 Wiser, R., Millstein, D., Mai, T., Macknick, J., Carpenter, A., Cohen, S., Cole, W., Frew, B. and Heath, G., 2016. The environmental and public health benefits of achieving high penetrations of solar energy in the United States. *Energy*, 113, pp.472-486. See also https://www.energy.gov/eere/solar/downloads/environmental-and-public-health-benefits-achieving-high-penetration-solar

21 Pearce, J. and Lau, A., 2002, January. Net energy analysis for sustainable energy production from silicon based solar cells. In *ASME Solar 2002: International Solar Energy Conference* (pp. 181-186). American Society of Mechanical Engineers. http://alpha.chem.umb.edu/chemistry/ch471/evans%20files/Net_Energy%20solar%20cells.pdf

possible to replace fossil fuels.[22] **PV is made from readily available, earth-abundant materials**. The vast majority of available PV modules use silicon as their main component.[23] This is good because there is plenty of silicon – it is the second most abundant element in the Earth's crust.[24] The silicon cells manufactured from one ton of sand produce as much electricity as burning 500,000 tons of coal.[25] Alternatively, conventional fuel sources have created an array of environmental and human health problems, including carbon dioxide emissions[26] leading to climate change,[27] acid rain,[28] air pollution,[29] smog,[30] water pollution,[31] rapidly-filling waste disposal sites,[32] and destruction of habitat from oil spills and accidents.[33] In the U.S., replacing coal-fired

22   Kenny, R., Law, C. and Pearce, J.M., 2010. Towards real energy economics: energy policy driven by life-cycle carbon emission. *Energy Policy*, 38(4), pp.1969-1978. http://mtu.academia.edu/JoshuaPearce/Papers/1540226/Towards_real_energy_economics_Energy_policy_driven_by_life-cycle_carbon_emission

23   Utility solar photovoltaic capacity is dominated by crystalline silicon panel technology. US EIA. 2017. https://www.eia.gov/todayinenergy/detail.php?id=34112

24   10 Most Abundant Elements in the Earth's Crust, Source: *CRC Handbook of Chemistry and Physics, 77th Edition* https://education.jlab.org/glossary/abund_ele.html

25   Williams, B.W., 2006. Principles and Elements of Power Electronics. *Devices, Drivers, Applications and Passive Components*. University of Strathclyde Glasgow

26   Weisser, D. A guide to life-cycle greenhouse gas (GHG) emissions from electric supply technologies. *Energy 2007*; 32(9): 1543-1559.

27   Change, Intergovernmental Panel On Climate. "IPCC." *Climate change* (2014).

28   Epstein, P.R., Buonocore, J.J., Eckerle, K., Hendryx, M., Stout Iii, B.M., Heinberg, R., Clapp, R.W., May, B., Reinhart, N.L., Ahern, M.M. and Doshi, S.K., 2011. Full cost accounting for the life cycle of coal. *Annals of the New York Academy of Sciences, 1219(1)*, pp.73-98.

29   Kampa, M, Castanas, E. Human health effects of air pollution. Environmental Pollution 2008;151(2): 362-367. | Curtis, L, Rea, W, Smith-Willis, P, Fenyves, E, Pan, Y. Adverse health effects of outdoor air pollutants. Environment International 2006;32(6): 815-830. | Gaffney, J, Marley, N. The impacts of combustion emissions on air quality and climate - from coal to biofuels and beyond. Atmospheric Environment 2009;43(1): 23-36. | Smith, K, Frumkin, H, Balakrishnan, K, Butler, C, Chafe, Z, Fairlie, I, Kinney, P, Kjellstrom, | T, Mauzerall, D, McKone, T, McMichael, A, Schneider, M. Energy and human health. Annual Review of Public Health 2013;34:159-188. | Finkelman, R, Orem, W, Castranova, V, Tatu, C, Belkin, H, Zheng, B, Lerch, H, Maharaj, | S, Bates, A. Health impacts of coal and coal use: possible solutions. International Journal of Coal Geology 2002;50(1-4); 425-443.

30   Zhang, J.J. and Samet, J.M., 2015. Chinese haze versus Western smog: lessons learned. *Journal of thoracic disease*, 7(1), p.3.

31   Winchester, J.W. and Nifong, G.D., 1971. Water pollution in Lake Michigan by trace elements from pollution aerosol fallout. *Water, Air, and Soil Pollution, 1(1)*, pp.50-64.

32   Komnitsas, K., Paspaliaris, I., Zilberchmidt, M. and Groudev, S., 2001. Environmental impacts at coal waste disposal sites-efficiency of desulfurization technologies. *Global Nest: the International Journal, 3(2)*, pp.135-142.

33   Head, I.M. and Swannell, R.P., 1999. Bioremediation of petroleum hydrocarbon contaminants in marine habitats. *Current opinion in Biotechnology, 10(3)*, pp.234-239.

power with solar power could save about 52,000 lives each year from reduction in air pollution alone, as seen in Figure 3.7.[34]

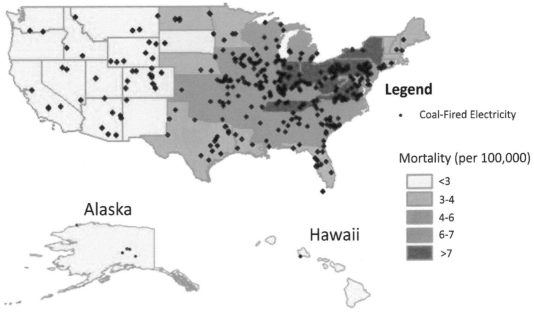

Legend

• Coal-Fired Electricity

Mortality (per 100,000)

	<3
	3-4
	4-6
	6-7
	>7

Alaska

Hawaii

Figure 3.7

Coal fired electricity facilities located in the U.S. and the annual mortality due to coal emissions per 100,000 people in each U.S. state. If solar were to replace coal, these premature deaths would be eliminated.[35]

⁑ **Sunlight is free**, which means there is no fuel cost for PV. Since there is no conventional fuel source, there is no required expenditure on the purchasing, storing, or transporting of the fuel itself. PV thus makes solar energy abundant. **PV moves us towards[36] a zero marginal cost society.[37]**

---

34 Prehoda, E.W. and Pearce, J.M., 2017. Potential lives saved by replacing coal with solar photovoltaic electricity production in the US. *Renewable and Sustainable Energy Reviews, 80*, pp.710-715. https://www.academia.edu/33288631/Potential_Lives_Saved_by_Replacing_Coal_with_Solar_Photovoltaic_Electricity_Production_in_the

35 ibid

36 Sawa. T. Toward a zero marginal cost society. Japan Times. https://www.japantimes.co.jp/opinion/2018/01/24/commentary/world-commentary/toward-zero-marginal-cost-society/

37 Rifkin, J., 2014. The zero marginal cost society: The internet of things, the collaborative commons, and the eclipse of capitalism. St. Martin's Press.

✳ PV costs have dropped precipitously,[38, 39] (as we will discuss in the next section), and now solar and wind are the least expensive electricity sources in most of the world.[40]

✳ **PV panels have extremely low operating and maintenance costs.**[41] For the small systems described in this book, this cost is negligible compared to costs of other renewable energy systems and conventional power plants. PV systems, in general, simply do not require frequent inspection or maintenance of any kind.[42]

✳ **PV works everywhere the sun shines**, which is most places that people live and work.[43] It thus makes it a largely geographically agnostic technology, which can reduce resource wars[44] and reduce threats to peace.[45]

✳ PV systems are **flexible** and can be designed for a variety of applications and operational requirements and can be used for either centralized or distributed power generation. **PV is ideally suited for distributed generation.**[46] Distributed generation is at the center of the future of smart grids and smart energy networks. In this book we focus on the latter – PV systems you can build yourself to fix your own problems. In some cases, PV systems can even be mobile/transportable and physically flexible[47] to be applicable to many types of situations.

---

38 Kavlak, G., McNerney, J. and Trancik, J.E., 2018. Evaluating the causes of cost reduction in photovoltaic modules. *Energy Policy*, *123*, pp.700-710.

39 Runyon, J. US Residential and Commercial Installed Solar PV Prices Still Dropping. Renewable Energy World. 2018. https://www.renewableenergyworld.com/articles/2018/09/us-residential-and-commercial-installed-solar-pv-prices-still-dropping.html

40 Berke, J. One simple chart shows why an energy revolution is coming — and who is likely to come out on top. Business Insider. 2018. https://www.businessinsider.com/solar-power-cost-decrease-2018-5

41 Gielen, D., 2012. Renewable energy technologies: cost analysis series. *Sol Photovolt*, *1*(1). Irena.

42 Some batteries do require maintenance and panels may require cleaning depending on the location.

43 Build Direct. 2016. Where Does Solar Energy Work Best? Everywhere https://www.builddirect.com/blog/where-does-solar-energy-work-best-everywhere/

44 Klare, M., 2001. Resource wars: the new landscape of global conflict. Metropolitan Books.

45 Joshua M. Pearce, "Reducing the Threat of a Nuclear Iran with Photovoltaic Technology: The Generous Solar Option", *Peace Studies Journal 8*(1), pp. 50-54 (2015). https://www.academia.edu/19646204/Reducing_the_Threat_of_a_Nuclear_Iran_with_Photovoltaic_Technology_The_Generous_Solar_Option

46 Caamaño-Martín, E., Laukamp, H., Jantsch, M., Erge, T., Thornycroft, J., De Moor, H., Cobben, S., Suna, D. and Gaiddon, B., 2008. Interaction between photovoltaic distributed generation and electricity networks. *Progress in Photovoltaics: research and applications, 16*(7), pp.629-643.

47 Engineering.com 2016. Flexible Solar Cells with Flexible Applications https://www.engineering.com/ElectronicsDesign/ElectronicsDesignArticles/ArticleID/12683/Flexible-Solar-Cells-with-Flexible-Applications.aspx

* **PV systems are highly reliable** and have no moving parts.[48] Even under the harshest of conditions, PV systems maintain electrical power supply. In comparison, conventional technologies often fail to supply power in the most critical of times.

* **PV modules are very durable and have long service lifetimes.** In general, modules carry a warranty of 80% of their rated power for 20 or more years. Thus, the worst case is an expected 1% decrease in performance per year. There have been several studies showing that in the real world, there is even less degradation than this – at around 0.2%-0.5%/year,[49, 50] PV panels can last up to thirty years or longer.[51]

* **PV panels are completely silent**, which reduces noise pollution if they offset generators. This also makes PV a superior solution for residential and other applications - like libraries and schools where silence is an asset - that are currently powered with a diesel generator (it should be noted that inverters do produce a modest hum).[52]

* **PV systems are modular and easily expandable.** Unlike conventional power systems, modules may be added to photovoltaic systems to increase available power. Recent improvements in microinverters and DC optimizers make this easier than ever before.[53]

* **PV systems are safe.**[54] As PV systems do not require the use of combustible fuels, they are very safe when properly designed and installed.

---

48 Except for tracking systems, which we do not recommend. PV panels have no mechanically moving parts; consequently they have far less breakages or require less maintenance than other energy generating technologies (e.g. generators or wind turbines).

49 A. Realini, MTBF - PVm, Mean Time Before Failure of Photovoltaic modules, Final report, June 2003, pp. 1-58.

50 D. Chianese, A. Realini, N. Cereghetti, S. Rezzonico, E. Bura, G. Friesen, Analysis of Weather c-Si PV Modules, Proc. *3rd World Conference on Photovoltaic Solar Energy Conversion*, 2003.

51 Other components of some PV systems, such as the battery, have much shorter life spans and may need to be replaced after several years of use.

52 Noise-free solar panels. RGS. 2015. https://rgsenergy.com/how-solar-panels-work/noise-free-solar-panels/

53 Comparing microinverters vs. power optimizers. Energy Sage. 2019. https://www.energysage.com/solar/101/microinverters-vs-power-optimizers/

54 Solar panel safety: how safe are solar panels? Energy Sage. https://news.energysage.com/solar-panel-safety-need-know/

✳ **PV can be used to shave peak loads.**[55] Solar energy availability coincides with energy needs for cooling. Hot, sunny summer days are when air conditioning (AC) loads are high and PV panels are churning out electrical energy to feed them. Thus, PV systems provide an effective solution to energy demand peaks – especially in hot summer months in regions where energy demand is high. This benefits the electric grid as well because it lowers conventional electricity market prices due to reduced peak demand.

✳ **PV provides electricity independence.**[56] PV systems can be designed to operate independent of the electric grid systems. This is a large advantage for rural communities in nations lacking basic infrastructure as well as for those that live too far from the conventional grid.

✳ **PV works great at high altitudes.**[57] When using solar energy, power output is optimized at higher elevations. This is very advantageous for high altitude, isolated communities where diesel generators must be de-rated due to the loss in efficiency and power output.

✳ Having grid-tied PV systems also helps improve the grid. **PV can improve grid reliability, which is of use to electric utilities** by providing voltage regulation, dynamic control, inverter ride through, and improved power quality.[58] Small-scale decentralized power stations reduce the possibility of power outages, which are often frequent on the electric grid. Similarly, detailed studies of electric outages on

---

55 Yang, Y., Li, H., Aichhorn, A., Zheng, J. and Greenleaf, M., 2014. Sizing strategy of distributed battery storage system with high penetration of photovoltaic for voltage regulation and peak load shaving. *IEEE Transactions on Smart Grid, 5*(2), pp.982-991. | Chowdhury, B.H. and Rahman, S., 1988. Analysis of interrelationships between photovoltaic power and battery storage for electric utility load management. *IEEE Transactions on Power Systems, 3*(3), pp.900-907. | Ortjohann, E. and Omari, O.A., 2002, May. Peak load shaving in conventional electrical grids by small photovoltaic systems in sunny regions. In Photovoltaic Specialists Conference, 2002. *Conference Record of the Twenty-Ninth IEEE* (pp. 1634-1637). IEEE.

56 Achieving energy independence through solar PV and battery storage. *Energy 2017.* https://www.energydigital.com/renewable-energy/achieving-energy-independence-through-solar-pv-and-battery-storage

57 Photovoltaic Power Plants Located in High Altitudes - Some Case Studies http://www.pvresources.com/en/pvpowerplants/highaltitudes.php

58 PV Generation and Its Effect on Utilities. Solar Professional 2013. https://solarprofessional.com/articles/design-installation/pv-generation-and-its-effect-on-utilities?v=disable_pagination&nopaging=1

U.S. military bases show **PV-powered micro-grids can improve national security.**[59]

✳ **PV can provide a valuable price hedge** from variably-priced fossil fuels because solar energy use is free and renewable.[60]

✳ **Distributed PV reduces costs for grid improvements** because of the avoided costs of new transmission and distribution infrastructure to manage electricity delivery from centralized power plants.[61] This also reduces the costs to build, operate, and maintain fossil fuel generating plants, which are greater costs than that of solar electricity.

✳ **PV offers reduced liability for power generators.** Many traditional sources of power have substantial externalities, and there are a growing number of lawsuits pending to compensate victims of these externalities. Thus, PV reduces future costs of mitigating the environmental impacts[62] of fossil fuels like coal[63] and nuclear[64] generation technologies.

✳ **PV also enhances tax revenues associated with local job creation,**[65] which is higher for solar than conventional power generation.[66] Solar creates both

59 Prehoda, E.W., Schelly, C. and Pearce, J.M., 2017. US strategic solar photovoltaic-powered microgrid deployment for enhanced national security. *Renewable and Sustainable Energy Reviews*, 78, pp.167-175. https://www.academia.edu/32808527/U.S._strategic_solar_photovoltaic-powered_microgrid_deployment_for_enhanced_national_security

60 Cooper, C., 2008. A national renewable portfolio standard: politically correct or just plain correct?. *The Electricity Journal*, 21(5), pp.9-17.

61 Minnesota Values Solar Generation with New "Value of Solar" Tariff. NREL. 2014. https://www.nrel.gov/state-local-tribal/blog/posts/vos-series-minnesota.html

62 Allen, M., 2003. Liability for climate change. *Nature*, 421(6926), p.891.

63 Heidari, N. and Pearce, J.M., 2016. A review of greenhouse gas emission liabilities as the value of renewable energy for mitigating lawsuits for climate change related damages. *Renewable and sustainable energy reviews*, 55, pp.899-908. https://www.academia.edu/19418589/A_Review_of_Greenhouse_Gas_Emission_Liabilities_as_the_Value_of_Renewable_Energy_for_Mitigating_Lawsuits_for_Climate_Change_Related_Damages

64 Laureto, J.J. and Pearce, J.M., 2016. Nuclear insurance subsidies cost from post-Fukushima accounting based on media sources. *Sustainability*, 8(12), p.1301. https://www.mdpi.com/2071-1050/8/12/1301/html

65 New Data Shows Solar Energy Creates More Jobs in America Than Any Other Industry Good for the environment, good for jobs. Futurism. 2018. https://futurism.com/new-data-shows-solar-energy-creates-more-jobs-america-other-industry

66 Lehr, U., Lutz, C. and Edler, D., 2012. Green jobs? Economic impacts of renewable energy in Germany. *Energy Policy*, 47, pp.358-364.

more jobs and higher paying jobs than fossil fuel-related employment.[67] For example, revenue generation for the Canadian government by supporting PV manufacturing is so substantial that it would be profitable to give solar manufacturing plants away for free.[68]

Now PV sounds pretty good…and it is, but it is not a perfect technology, and it does have some disadvantages, which you should be aware of when designing a system in order to be able to minimize or overcome them. First, PV does not work well when the sun does not shine (cloudy days, rainy/snowy weather, and at night) and it makes PV generation difficult to predict (i.e. variable or uncertain). PV can, of course, be coupled with storage like batteries to counteract this issue and provide continuous power. However, adding storage or a backup system increases the costs and complexity. Second, the electricity PV produces is direct electricity (DC), which is great if that is your load, but if you need alternating electricity (AC) power, you will need an inverter, which again increases costs and complexity. Large systems providing a lot of power, and the relatively small energy density of PV, demand relatively large areas for deployment. This can cause a conflict with other land uses like agriculture. Luckily, both agrivoltaics, [69, 70] (interplanting crops between PV rows) and floatovoltaics[71]/aquavoltaics[72] (deploying PV over water surfaces) provide good solutions to this problem with several synergistic benefits like increased

67 Louie, E.P. and Pearce, J.M., 2016. Retraining investment for US transition from coal to solar photovoltaic employment. *Energy Economics, 57*, pp.295-302. https://www.academia.edu/26372861/Retraining_Investment_for_U.S._Transition_from_Coal_to_Solar_Photovoltaic_Employment

68 Branker, K. and Pearce, J.M., 2010. Financial return for government support of large-scale thin-film solar photovoltaic manufacturing in Canada. *Energy Policy, 38*(8), pp.4291-4303. http://mtu.academia.edu/JoshuaPearce/Papers/1540699/Financial_Return_for_Government_Support_Financial_Return_for_Government_Support_of_Large-Scale_Thin-Film_Solar_Photovoltaic_Manufacturing_in_Canada

69 Dinesh, H. and Pearce, J.M., 2016. The potential of agrivoltaic systems. *Renewable and Sustainable Energy Reviews, 54*, pp.299-308. https://www.academia.edu/18406368/The_potential_of_agrivoltaic_systems

70 Dupraz, C., Marrou, H., Talbot, G., Dufour, L., Nogier, A. and Ferard, Y., 2011. Combining solar photovoltaic panels and food crops for optimising land use: towards new agrivoltaic schemes. *Renewable energy, 36*(10), pp.2725-2732.

71 Kumar, N.M., Kanchikere, J. and Mallikarjun, P., 2018. Floatovoltaics: Towards improved energy efficiency, land and water management. *International Journal of Civil Engineering and Technology, 9*, pp.1089-1096.

72 Pringle, A.M., Handler, R.M. and Pearce, J.M., 2017. Aquavoltaics: Synergies for dual use of water area for solar photovoltaic electricity generation and aquaculture. *Renewable and Sustainable Energy Reviews, 80*, pp.572-584. https://www.academia.edu/33327275/Aquavoltaics_Synergies_for_Dual_Use_of_Water_Area_for_Solar_Photovoltaic_Electricity_Generation_and_Aquaculture

income per acre and water conservation.[73] Finally, although PV panels have negligible maintenance or operating costs, they are generally made with glass front panels and are therefore fragile and can be damaged relatively easily. This is primarily a concern during transportation and installation. Once installed, breakage is extremely uncommon. Although recycling PV remains challenging and limited.[74] Insurance is one method to overcome this risk.

As you can see, the disadvantages are overwhelmed by the positive benefits of PV. There is little doubt in our minds that solar will power the significant part of society in the future![75] PV represents one of the most promising means of maintaining our energy-intensive standard of living while not contributing to global warming and pollution.

Solar photovoltaics is the future of energy!

73 Hayibo, K.S., Mayville, P., Kailey, R.K. and Pearce, J.M., 2020. Water Conservation Potential of Self-Funded Foam-Based Flexible Surface-Mounted Floatovoltaics. *Energies, 13*(23), p.6285. https://doi.org/10.3390/en13236285

74 McDonald, N. C., & Pearce, J. M. (2010). Producer responsibility and recycling solar photovoltaic modules. *Energy Policy, 38*(11), 7041-7047. https://www.academia.edu/1485052/Producer_Responsibility_and_Recycling_Solar_Photovoltaic_Modules

75 Pearce, J.M., 2002. Photovoltaics—a path to sustainable futures. *Futures, 34*(7), pp.663-674. http://www.academia.edu/download/12060956/futures.pdf

3.3. Costs

In the "old-days," solar photovoltaic devices were the most expensive component of a system, and designers would go to great lengths to try to stretch the performance of every module as much as possible. For instance, PV system designers would use sophisticated multiple axis trackers to ensure that the PV was always pointed right at the sun so that the high costs of the PV panels could be recouped. Luckily, solar prices have dropped like a rock,[76] so a lot of the old rules of thumb about solar no longer apply. In fact, the most expensive component of a PV system is often now the racking – the rather mundane mechanical structure holding the module off the ground or roof.

The continuous drop in PV module prices as the global PV industry has matured and grown is called a **"learning curve".**[77] This learning has resulted in continuous and aggressive reduction in the costs of solar modules[78] and is sometimes called **Swanson's law**. Richard Swanson, the founder of SunPower Corporation (a PV panel manufacturer), observed that the price of PV modules tends to drop 20% for every doubling of cumulative shipped volume, as can be seen in Figure 3.8. Figure 3.8 shows that the actual decrease in PV costs[79] are actually a bit better than a 20% drop for every doubling from *The International Technology Roadmap for Photovoltaic.*[80] These costs are rapidly moving; for example, the lowest multi-silicon PV module price in September 2021 is 21.6 cents per watt (18-9-2021).

76 Feldman, D.; Barbose, G.; Margolis, R.; Wiser, R.; Darghout, N.; Goodrich, A. Photovoltaic (PV) Pricing Trends: Historical, Recent, and Near-term Projections; Technical Report; Lawrence Berkeley National Lab.: Berkeley, CA, USA, 2012.

77 Mauleón, I. Photovoltaic learning rate estimation: Issues and implications. *Renew. Sustain. Energy Rev. 2016, 65,* 507–524.

78 Barbose, G.L.; Darghouth, N.R.; Millstein, D.; LaCommare, K.; DiSanti, N.; Widiss, R. Tracking the Sun 10: The Installed Price of Residential and Non-Residential Photovoltaic Systems in the United States. Available online: https://emp.lbl.gov/publications/tracking-sun-10-installed-price (accessed on 25 July 2018).

79 Energy Trend. Spot prices. https://www.energytrend.com/solar-price.html

80 International Technology Roadmap for Photovoltaic http://www.itrpv.net/Reports/Downloads/2017/

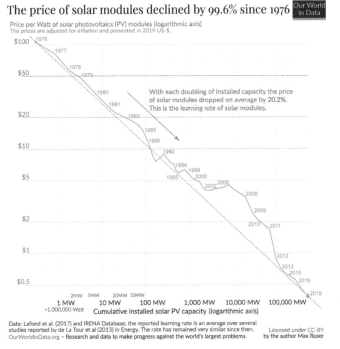

The price of solar modules declined by 99.6% since 1976

Price per Watt of solar photovoltaics (PV) modules (logarithmic axis)
The prices are adjusted for inflation and presented in 2019 US-$.

With each doubling of installed capacity the price of solar modules dropped on average by 20.2%. This is the *learning rate* of solar modules.

Cumulative installed solar PV capacity (logarithmic axis)

Data: Lafond et al. (2017) and IRENA Database; the reported learning rate is an average over several studies reported by de La Tour et al (2013) in Energy. The rate has remained very similar since then. OurWorldinData.org – Research and data to make progress against the world's largest problems.

Licensed under CC-BY by the author Max Roser

Figure 3.8

Swanson's law is a 20% decrease in price for every doubling of cumulative installed photovoltaics (blue dotted line). The solid orange line shows actual worldwide module shipments vs. average module price, from 1976 ($104/Wp) to 2019 ($0.36/Wp). Prices are in 2019 dollars. https://commons.wikimedia.org/wiki/File:Solar-pv-prices-vs-cumulative-capacity.png CC-BY by Max Roser

Log scale graphs, like Figure 3.8, make beautiful straight lines for exponential changes but can be difficult to understand if you are not used to looking at them. However, you can also compare the astronomical price drop of PV on a linear scale as shown in Figure 3.9. The global PV installations in 2018 passed 100 GW (up to 104 GW),[81] and the International Renewable Energy Agency (IRENA) predicts that the prices will continue to fall by 60% in the next decade.[82]

81   Global Solar PV Installations to Surpass 104GW in 2018. GTM. 2018. https://www.greentechmedia.com/articles/read/global-solar-pv-installations-to-surpass-104-gw-in-2018#gs.SqrNvRCJ

82   Reuters. Solar Costs to Fall Further, Powering Global Demand-Irena. Available online: https://www.reuters.com/article/singapore-energy-solar/solar-costs-to-fall-further-powering-global-demand-irena-idUSL4N1MY2F8 (accessed on 5 March 2018).

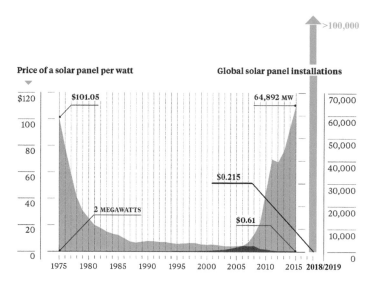

Figure 3.9
The radical drop in prices from 1975 to 2019 as the global solar PV installations exploded. Graph adapted from CleanTechnica[83] with data from Bloomberg, Earth Policy Institute, GTM, and Energy Trend.

So, we know the cost of PV modules has dropped, but does that really make PV affordable for you?

In short, the answer is – Yes!

To know for sure, you need to calculate what the cost of electricity is based on the cost of the PV system. You do this by dividing that initial, fully installed capital cost by the electricity that is produced by the PV over its lifetime. This is called the **levelized cost of electricity (LCOE).**[84] The calculation is a little tricky because to do it properly, you need to use discounting values in the future (We will cover this in detail in Section 8.3, if you are interested). **The shortcut to the answer that matters for your wallet**

83 Shahan, Z. Solar Panel Prices Continue Falling Quicker Than Expected (#CleanTechnica Exclusive) CleanTechnica. 2018. https://cleantechnica.com/2018/02/11/solar-panel-prices-continue-falling-quicker-expected-cleantechnica-exclusive/

84 Branker, K., Pathak, M.J.M. and Pearce, J.M., 2011. A review of solar photovoltaic levelized cost of electricity. *Renewable and sustainable energy reviews*, *15*(9), pp.4470-4482. https://www.academia.edu/1484968/A_Review_of_Solar_Photovoltaic_Levelized_Cost_of_Electricity

is that at current prices, PV installations provide an LCOE lower than residential electricity prices from the grid, and at utility scales, PV is cost competitive with all conventional sources[85] **in many regions throughout the world.** The actual installed cost varies quite dramatically depending on where you live because of tariffs, taxes, regulations, and labor costs. Again, not so long ago, PV modules made up about half of the cost of a system. Now, the majority of the cost is soft (e.g., labor) for small-medium sized PV systems on the order of a few kW. We can see that the prices for house- and building-sized systems is only a few dollars per watt throughout most of the world, as seen in Figure 3.10.[86] The costs range from over \$3/W in the U.S.A. to just a \$0.70/W in China at the residential level.

Notes: Installed prices for countries other than the USA are from the International Renewable Energy Agency (IRENA)'s "Renewable Power Generation Costs in 2020" report and are derived from IRENA's Renewable Cost Database. For the Non-Residential sector, data from IRENA generally refer to systems up to 500 kW in size, and thus encompass both the Small and some portion of the Large Non-Residential segment used within Tracking the Sun.

Figure 3.10

Comparing installed PV systems' pricing for residential and non-residential systems in various selected countries (Tracking the Sun).[87]

85   Safi, M. Indian Solar Power Prices Hit Record Low, Undercutting Fossil Fuels. Available online: http://www.theguardian.com/environment/2017/may/10/indian-solar-power-prices-hit-record-low-undercutting-fossil-fuels (accessed on 5 March 2018).

86   Berkeley Lab's Tracking the Sun report visited 10-10-2021. https://emp.lbl.gov/tracking-the-sun

87   ibid

Conventional thinking is that the larger a system is, the lower the installed cost per W. You can see how people get this idea by looking at Figure 3.11. If we look, for example, at residential systems in 2016, they were just under $4/W, while the large systems were just over $2/W. However, conventional thinking fails to account for the exceptionally low-costs of PV systems at the microscale, which is the topic of this book. If you do the installation yourself, that cuts the major cost for most PV systems right off the top. If you can build the rack or do the wiring yourself, that also radically reduces the costs. In addition, for small systems, the cost of electricity from the grid may not be what you should compare it to – it might be the cost of batteries (which are far more expensive than electricity from the grid)[88] or the cost of running a generator or extending the grid to your site.

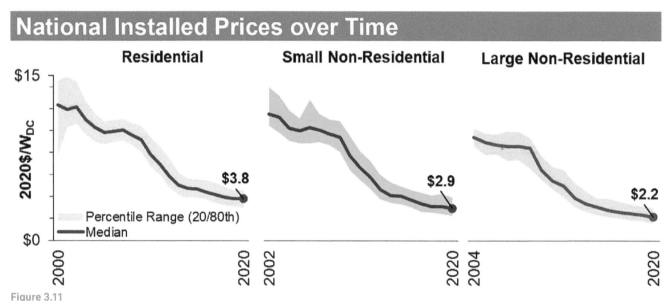

National Installed Prices over Time

Figure 3.11

Residential, non-residential small and large system price declines in the US-- Tracking the Sun.[89]

88   Rhett Allain. Are Expensive Batteries Worth the Extra Cost? Wired. https://www.wired.com/2012/01/are-expensive-batteries-worth-the-extra-cost/

89   Berkeley Lab's Tracking the Sun report–Distributed Solar 2021 Data Update. https://emp.lbl.gov/tracking-the-sun

Regardless of the size of your PV system, financing plays an enormous role in the profitability of solar projects, and new methods of solar financing including third party,[90] peer-to-peer,[91] securitization,[92] credit trading,[93] and government policies to reduce pollution,[94, 95] have become available and widespread (e.g., Sunrun, Solar City/Tesla, etc. in the U.S.) and have increased access to PV systems for everyone.[96] For the PV industry to expand electricity market share into the future,[97] a lot more solar needs to be deployed. That is where you can come in. Historically, PV systems costs were lowered due to decreased module prices. However, today, balance of systems (BOS) and installation costs make up a greater fraction of a system's costs; therefore, PV module prices are less important for the overall system's cost. For example, among projects covered in Tracking the Sun 10, median module efficiencies grew from 12.7% to 17.3% in the years 2002 to 2016, which enabled the average systems' sizes to more than double, while driving a $1/W system costs decline.[98]

PV systems at the microscale (<1 kW) can push the system's cost much lower because labor and some BOS costs can be eliminated. For example, the average person does not have easy access to capital/financing to install a PV system able to meet their aggregate annual electric needs. One method to overcome this challenge is to allow "plug-and-play

90 Drury, E.; Miller, M.; Macal, C.M.; Graziano, D.J.; Heimiller, D.; Ozik, J.; Perry, T.D., IV. The transformation of southern California's residential photovoltaics market through third-party ownership. *Energy Policy 2012, 42*, 681–690.

91 Branker, K.; Shackles, E.; Pearce, J.M. Peer-to-peer financing mechanisms to accelerate renewable energy deployment. *J. Sustain. Financ. Invest. 2011, 1*, 138–155. http://mtu.academia.edu/JoshuaPearce/Papers/1540666/Peer-to-Peer_Financing_Mechanisms_to_Accelerate_Renewable_Energy_Deployment

92 Alafita, T.; Pearce, J.M. Securitization of residential solar photovoltaic assets: Costs, risks and uncertainty. *Energy Policy 2014, 67*, 488–498. https://www.academia.edu/6400583/Securitization_of_residential_solar_photovoltaic_assets_Costs_risks_and_uncertainty

93 Hede, S.; Nunes, M.J.L.; Ferreira, P. Credits trading mechanism for corporate social responsibility: An empirically grounded framework. *Int. J. Technol. Learn. Innov. Dev. 2014, 7*, 49–92.

94 Overholm, H. Spreading the rooftop revolution: What policies enable solar-as-a-service? *Energy Policy 2015, 84*, 69–79.

95 Ameli, N.; Kammen, D.M. Innovations in financing that drive cost parity for long-term electricity sustainability: An assessment of Italy, Europe's fastest growing solar photovoltaic market. *Energy Sustain. Dev. 2014, 19*, 130–137.

96 Coughlin, J.; Cory, K.S. Solar Photovoltaic Financing: Residential Sector Deployment; Technical Report; National Renewable Energy Laboratory: Golden, CO, USA, 2009.

97 International Energy Agency. *Renewables 2017*. Available online: https://www.iea.org/publications/renewables2017/ (accessed on 5 March 2018).

98 Barbose, G.L.; Darghouth, N.R.; Millstein, D.; LaCommare, K.; DiSanti, N.; Widiss, R. Tracking the Sun 10: The Installed Price of Residential and Non-Residential Photovoltaic Systems in the United States. Available online: https://emp.lbl.gov/publications/tracking-sun-10-installed-price (accessed on 25 July 2018).

solar," which is defined as a fully inclusive, off-the-shelf PV system (normally consisting of a PV module and microinverter), which a prosumer (producing consumer) can install by plugging it into an electric outlet and avoiding the need for significant permitting, inspection, and interconnection processes. Many advanced countries already allow plug-and-play solar (e.g., the U.K. and throughout some of the European Union). A recent study reviewed the relevant codes and standards from the National Electric Code in the U.S., local jurisdictions, and utilities for PV with a specific focus on plug-and-play solar. It found that commercially available microinverters and alternating current (AC) PV modules met technical and safety compliance to these standards.[99] If such plug-and-play systems were deployed across the U.S. in households with appropriate orientations and capital, it would save consumers over $13 billion/year.[100]

However, there are even less expensive micro-PV systems such as PV directly powering a water pump. In this case, if you do the work yourself, it is only the cost of the modules, modest wiring, and racking. For pretty much everywhere in the world, this is far less than $1/W, which is less than large system installs even in the most advanced economies.

99 Aishwarya S. Mundada, , Yuenyong Nilsiam , Joshua M. Pearce. A review of technical requirements for plug-and-play solar photovoltaic microinverter systems in the United States. *Solar Energy 135*, (2016), pp. 455–470. doi: 10.1016/j.solener.2016.06.002 https://www.academia.edu/26379506/A_Review_of_Technical_Requirements_for_Plug-and-Play_Solar_Photovoltaic_Microinverter_Systems_in_the_United_States

100 Aishwarya S. Mundada, Emily W. Prehoda, Joshua M. Pearce. U.S. market for solar photovoltaic plug-and-play systems. *Renewable Energy. 103* (2017) pp. 255–264. DOI:10.1016/j.renene.2016.11.034 https://www.academia.edu/29941674/U.S._Market_for_Solar_Photovoltaic_Plug-and-Play_Systems

3.4. Photovoltaic effect

The **photovoltaic effect** occurs when two materials in close contact are exposed to light, producing an electrical voltage. **Solar photovoltaic (PV) devices thus convert sunlight directly into electricity.** Understanding that PV converts light to electricity is enough to use them. If you want to understand the magic, see the side bar.

How solar cells work.

Normally, electrons are held tightly to the nuclear core of an atom. The further the electrons move from the nucleus, the more energetic they are. When packets of light, called photons, strike a semiconductor like silicon, they provide enough energy for electrons to break free of their host atoms and move around the semiconductor. Eventually, the electrons lose energy (get tired) and fall back to the ground states and become shackled to a nucleus again. If a semiconductor material is by itself, it is not very exciting, as nothing happens. However, if you put two dissimilar semiconductors next to each other – like one doped to extra electrons and one doped to have a shortage of electrons – you create a solar cell junction. Photon generated electrons cross the junction between the two differently doped semiconductors more easily one way than the other, which gives one side of the junction a negative charge and, therefore, a negative voltage with respect to the other side. This is the exact same charge separation you see in a battery, where one side is positive and one side is negative. The cool thing about a solar cell is that it will continue to provide a voltage and a current – electricity! – as long as it is illuminated.

An easy way to understand the photovoltaic effect of a solar cell is to consider children playing on a playground as electrons. In the beginning, both the electrons and the children are in the "ground state" – that is ok, but not very exciting (see Figure 3.12).

Figure 3.12
Children playing on a slide at the bottom are like electrons in the "ground state" without much energy.

Figure 3.13
Electrons in the "excited state" are like children playing on a slide at the top – they are excited because they have a lot of potential energy.

When light is shined on the semiconductor, the light energy can be absorbed and consumed and lift the electrons up to an "excited state." In the same way, children consume chemical energy stored in their bodies from eating food and can lift themselves up a ladder or rock wall to an excited state (see Figure 3.13). For both the electrons and the children in the excited state (at the top of the slide), there is energy that can be used. If there is no photovoltaic junction in the semiconductor, the electrons just fall back to the ground state. However, when a junction is present in a photovoltaic device, an electric field is generated and "tilts" the energy levels, which exert a force on the free electrons into an external electrical load like a light bulb or a computer where their excess energy can be dissipated. Similarly, when children are at the top of the play structure, they tend to move towards the slide because it is more exciting. As the children slide down the slide, they move like electrons in the junction, and they dissipate their excess energy as they reach the ground state again. The motion of the electrons, like the children, is only in one direction (no running up the slide!). For the electrons, this is a direct current (DC) just like a battery. Both the electrons and the kids can start the whole process over again as long as there is energy (light for the electrons in the solar cell and food energy stored in the children). If the sun goes down, then there is no energy to lift the electrons up, and if the children play too long and use up all their stored chemical energy, they get hungry and eventually lethargic and will no longer climb up to the top of the play structure.

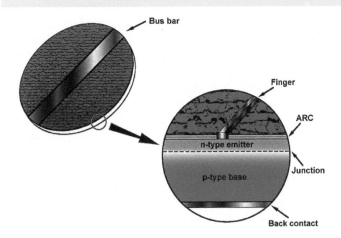

Figure 3.14
Details of the construction of a silicon solar cell.

A solar cell is a bit like a sandwich with semiconductor filling and bread slices made up of conductors. There is a metal conductor on the back of the solar cell. Electric current is generated in the semiconductor (blue area in Figure 3.14), which has the two doped layers that make up the p-n junction. To get light to the semiconductors, the front of the solar cell must be transparent to light. This is hard to do with metal, so solar cell designers have historically used thin metal strips (fingers) that let the light pass and carry a little bit of electricity generated near them to a busbar that can handle larger currents when all the electricity is collected from the fingers.

On top of the solar cell there is an antireflection coating, or ARC, which is used to cover the cell to minimize light reflection from the top surface. ARCs are made with a thin layer of dielectric material. When light comes in, you want to collect as much of it as possible, and if you can see light reflecting off the solar cell, the ARC is not good.

3.5. Types of panels

The U.S. National Renewable Energy Laboratory (NREL), which mostly funded Joshua's PhD research, maintains a database of the highest confirmed conversion efficiencies for turning light into electricity for research PV cells for a number of materials, from 1976 to the present. The NREL graph showing the efficiencies of all the technologies is shown in Figure 3.15.[101]

---

101 National Renewable Energy Laboratory - National Center for Photovoltaics https://www.energy.gov/eere/solar/downloads/research-cell-efficiency-records

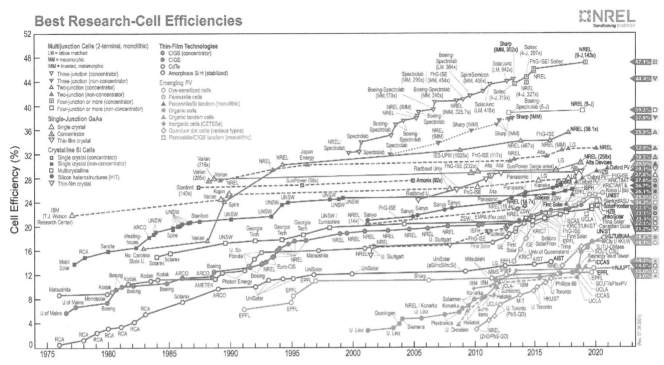

Figure 3.15
Solar cell efficiency records (NREL).

We should stress that the values shown in Figure 3.15 are record lab cells. It normally takes several years, maybe five to ten, before a record efficiency cell technology is transferred to the market. For example, although Panasonic has demonstrated **HIT** (heterojunction with thin intrinsic layer, which is a sandwich of small bandgap crystalline silicon between two layers of wide-bandgap amorphous silicon) solar cells with efficiency of over 25%, they are currently selling them at just under 20%.[102] This is, in part, because the records are over cells, not modules – and these cells normally have small areas. For example, although modules are about the size of a person with their arms stretched out, standard crystal semiconductor technology is about the size of your hand. The solar cells we work on in the lab for thin film experiments are generally only 2 mm in diameter or about

---

102 https://na.panasonic.com/us/energy-solutions/solar/hit-modules/n325k-photovoltaic-module-hitr-black-40mm

double the width of your fingernail. Scaling technologies to large areas normally results in a modest drop in efficiency with an additional drop moving up to the module level.

The chart in Figure 3.15 is arranged by the different families of semiconductors: (1) multijunction cells - purple, (2) single-junction gallium arsenide cells - purple, (3) crystalline silicon cells - blue, (4) thin film technologies - green, and (5) emerging photovoltaics - orange. Now, everyone knows that the higher the efficiency, the better, so it may be tempting to look only at the top of the list at the impressive efficiencies made up of multijunction cells. Also, it should be pointed out here that the efficiency of solar cells should not be directly compared to other technologies like a generator because the fuel for PV is free, while generator fuel accounts for the majority of the costs. In the future, multijunction cells that can convert a greater range of the solar flux into electricity may dominate; however, for now, the cells in groups 1 and 2 are primarily used for space and military applications because of their extreme costs. Group 5 emerging technologies are interesting because, as you can see from Figure 3.15, they are improving rapidly and may soon be commercialized.

The vast majority of the solar cells on the market are silicon based – crystalline, polycrystalline (also called multi-crystalline), or amorphous silicon (group 3 and group 4 in Figure 3.15: Solar cell efficiency records (NREL). We will compare the advantages and disadvantages of each. First, we will look at crystalline and polycrystalline silicon. Advantages include high efficiency (over 20%), very established technology, and stability. The technology for the crystalline silicon is the same technology that drives the semiconductor industry, so it is mature and established. The disadvantages for crystalline silicon technology include low absorption (you need a lot of material to absorb a sufficient amount of sunlight) and the need for a large amount of highly purified silicon starting material, which is expensive. Polycrystalline silicon is cheaper to produce, but the electric properties are not as good as pure crystalline silicon, so the efficiency is lower (usually in the mid to high teens). Today, the industry is moving mostly to pure crystalline silicon. You can see that the polycrystalline material is not as good as pure crystalline silicon because of the lines between crystals, as shown in Figure 3.16.

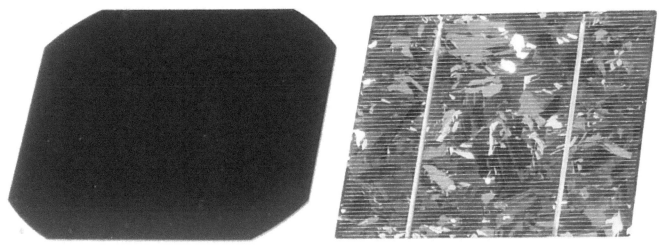

Figure 3.16
Comparison of a single crystalline silicon (left) solar cell ready for processing with a finished polycrystalline silicon solar cell (right).
https://commons.wikimedia.org/wiki/File:Comparison_solar_cell_poly-Si_vs_mono-Si.png CC-By-SA

Crystalline and polycrystalline silicon are the most common PV modules types and come in the widest variety of sizes. The standard size is quite large, as seen in Figure 3.17, where Joshua is adjusting an array that powers a research center in the Midwest of the U.S. The modules are about the size of a short adult and just small enough that a single person can manage them with minimal difficulty if done carefully.

Figure 3.17
A polycrystalline PV array that powers a research center in the Midwest U.S. that is used for cold weather testing.

The other kind of silicon solar cell material is amorphous silicon. It has a very high absorption coefficient compared to crystalline silicon, so you need only a very thin layer of it – only a few 100 nanometers (about a tenth of the thickness of your fingernail). To put the thinness of a thin film solar cell into perspective, consider the bag of sand shown in Figure 3.18. Sand is silicon dioxide. If you were to strip away the oxygen and use the remaining silicon in this bag to make amorphous silicon solar cells, you could cover half of the roof area of the typical house in Detroit, Michigan (not the sunniest place in the world), and these solar cells would provide all of the electricity needs for the notoriously energy-gluttonous American family living there.

Figure 3.18
Bag of sand.

In addition, thin film solar cell materials like amorphous silicon can be deposited on various materials - glass, metal, and even plastic. This allows for some interesting potential applications like flexible or peal-and-stick solar cells, as shown in Figure 3.19.

Figure 3.19
*A large roll of flexible peal-and-stick amorphous
silicon solar cell in Joshua's lab.*

It can also come as a sticker, such as the solar cell in the image, where all you do is peel off the backing and stick it to the building and it lasts for a very long time (30-year warranty).[103] The primary disadvantage of amorphous silicon is efficiency (around 10%) compared to the crystalline forms of silicon (around 20%). Finally, it degrades when light hits it, a phenomenon known as the Staebler-Wronski effect or SWE. All amorphous silicon solar cells sold now have warrantees for the degraded steady state, where SWE has stopped. For example, you might buy a solar cell with a warrantee of 10% efficiency. When you first install it, you will be happy to find that it will produce electricity at 11% efficiency. After a few months, it will degrade down to 10% and then stay there the rest of its lifetime.

---

103   We will talk more about the actual lifetime of a PV module in the economics section.

There are a few other thin film technologies to note, such as CIGS and CdTe. CIGS is short for copper indium gallium di-selenide or CuInGaSe2. CIGS solar cells have high efficiencies (by 2010, they had broken the 20% barrier) and high absorption, so you do not need a lot of material, similar to amorphous silicon. The disadvantages are that they have some instability problems, they are not as established as silicon (so, higher prices because of the smaller scale), and they have a CdS layer that contains toxic cadmium. This brings us to cadmium telluride or CdTe. These cells have high efficiencies, are thin, and they use a waste product (Cd) of the zinc, lead, and copper industries as the main absorber. Often, they are the lowest cost option, as one of the manufacturers (First Solar) has obtained a large scale. Their modules, however, are frameless and break easily, so they are normally used only for solar farms and handled only by professionals.

Figure 3.20
Frameless CdTe modules being tested with an alternative racking system. These modules were found to be too fragile for small-scale applications.

Other considerations when choosing your solar modules include framed versus frameless. Frameless modules are better for the environment because they eliminate all the embodied energy in the frame and are a good choice for solar farms in snowy areas[104] because they shed snow more easily, but they are more fragile, requiring careful handling and a special support for the racking. Several of the frameless CdTe modules we tested with alternative racking technology (see Figure 3.20) broke on the seven-mile car ride from campus to our outdoor testing facility after having been loaded in a pickup truck and driven over paved roads. We concluded they are too fragile for small-scale applications.

In the near future, you can expect most silicon-based PV panels to be made from black silicon. They operate the same way as traditional bluish panels but are better at capturing light because of etched nanostructures that make them look black and thus reduce the cost per watt.[105] The other major near-future area of PV technology trends is the use of bifacial panels, which accept light from both sides. They receive an energy output bump from reflected light (albedo) hitting the back of the module, which tends to increase energy output. It also helps melt snow off the front of the panels in the winter. For example, bifacial modules on a dual-axis tracker in Vermont produced 14% more electricity in a year than their monofacial counterparts and as much as 40% during the peak winter months.[106] Higher efficiencies will help lower costs, as less racking and wiring are needed if the number of panels is minimized. That said, for many applications, a full person-sized solar module of any type of material is unnecessary. Even very tiny PV modules can be used for all kinds of useful applications like providing light (Figure 3.21), or an array of three mini-modules can be used to power a laptop and a 3-D printer (Figure 3.22).[107]

104　Riley, D., Burnham, L., Walker, B. and Pearce, J.M., 2019, June. Differences in Snow Shedding in Photovoltaic Systems with Framed and Frameless Modules. In *2019 IEEE 46th Photovoltaic Specialists Conference (PVSC)* (pp. 0558-0561). IEEE. https://www. osti.gov/servlets/purl/1640734

105　Modanese, C., et al., 2018. Economic advantages of dry-etched black silicon in passivated emitter rear cell (PERC) photovoltaic manufacturing. *Energies, 11*(9), p.2337. https://doi.org/10.3390/en11092337

106　Burnham, et al. 2019, June. Performance of bifacial photovoltaic modules on a dual-axis tracker in a high-latitude, high-albedo environment. In *2019 IEEE 46th Photovoltaic Specialists Conference (PVSC)* (pp. 1320-1327). IEEE. https://www.osti.gov/servlets/ purl/1641282

107　Jephias Gwamuri, Dhiogo Franco, Khalid Y. Khan, Lucia Gauchia and Joshua M. Pearce. High-Efficiency Solar-Powered 3-D Printers for Sustainable Development. *Machines 2016, 4*(1), 3; https://doi.org/10.3390/machines4010003

Figure 3.21
School children from Kembu Primary School holding solar lights powered with mini-modules in Kenya. Image credit SolarAid. https://www.flickr.com/photos/solaraid/16748134826 CC-BY

Figure 3.22
Three mini crystalline silicon PV modules are used to power a laptop and mobile 3-D printer in the US.

3.6.   Insolation, full sun power, peak sun hours

No matter what part of the world you call home, on any given day, the solar radiation varies continuously from sunup to sundown, and it is also known to change spectrum (i.e., color of the light). Solar radiation also depends on cloud cover, sun position, and content and turbidity of the atmosphere. In addition, different types of solar cells respond to the spectrum differently and even respond differently to the color of light bouncing off the ground![108, 109] As you might suspect, making a perfect computer model of such a complex system can become a large mess with a lot of complicated formulas, but for the vast majority of PV systems, we can take shortcuts. Here, we will show you the shortcuts, which will be more than sufficient to get you to a reasonably close approximation of the electricity output for smaller PV systems.

The maximum **irradiance** (solar flux per unit area) is available at solar noon, which is defined as the midpoint, in time, between sunrise and sunset. **Insolation**, now commonly referred to as **irradiation** (and also known as peak sun hours), differs from irradiance because of the inclusion of time. Insolation is the amount of solar energy received on a given area over time, measured in kilowatt-hours per meter squared. Recall that a kilowatt-hour (kWh) is the way electricity is normally charged on an electric bill.

Insolation measured in $kWh/m^2/day$ is the equivalent to "peak sun hours" (which is in units of hours/day of full sun power). We know that in the early morning after sunrise and in the evening before sunset, we do not have as much solar power coming from the sun as we do at noon. Imagine if you could take all these time periods of dim sun hours and combine them to make full sun hours equivalent to a cloudless noon. This is a peak sun hour. "Peak sun hours" is defined as the equivalent number of hours per day, with solar irradiance equaling 1,000 W/m^2. This 1,000 W/m^2 is defined as the **full sun power**. Peak sun hours only make sense because PV panel power output is rated with a

108 Andrews, R.W. and Pearce, J.M., 2013. The effect of spectral albedo on amorphous silicon and crystalline silicon solar photovoltaic device performance. *Solar Energy, 91*, pp.233-241. https://www.academia.edu/3081684/The_effect_of_spectral_albedo_on_amorphous_silicon_and_crystalline_silicon_solar_photovoltaic_device_performance

109 Brennan, M.P., Abramase, A.L., Andrews, R.W. and Pearce, J.M., 2014. Effects of spectral albedo on solar photovoltaic devices. *Solar Energy Materials and Solar Cells, 124*, pp.111-116. https://www.academia.edu/6222506/Effects_of_spectral_albedo_on_solar_photovoltaic_devices

radiation level of 1,000 W/m². In other words, six peak sun hours means that the energy received during total daylight hours equals the energy that would have been received had the sun shone for six hours with an irradiance of 1,000 W/m². Many tables of solar data are often presented as an average daily value of peak sun hours (kWh/m²) for each month. Most places have between three and six peak sun hours on average per day throughout the entire year. In short, if you determine the number of peak sun hours you have for your area, it makes the math a little bit easier. You can get the basic idea of the irradiation over most of the world in kWh/m² per day and per year from this map. The daily sum with the legend is your "sun hours."

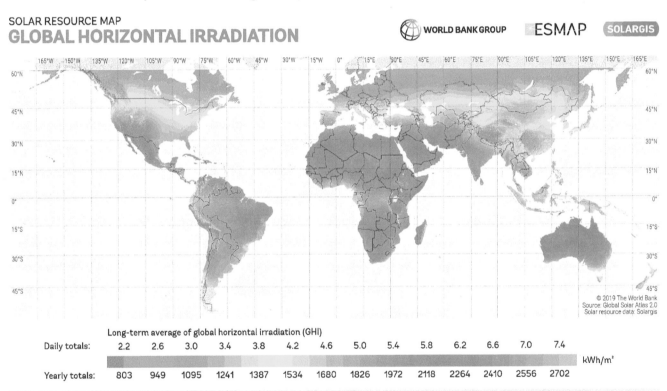

SOLAR RESOURCE MAP
GLOBAL HORIZONTAL IRRADIATION

WORLD BANK GROUP   ESMAP   SOLARGIS

© 2019 The World Bank
Source: Global Solar Atlas 2.0
Solar resource data: Solargis

Long-term average of global horizontal irradiation (GHI)

Daily totals:	2.2	2.6	3.0	3.4	3.8	4.2	4.6	5.0	5.4	5.8	6.2	6.6	7.0	7.4	kWh/m²
Yearly totals:	803	949	1095	1241	1387	1534	1680	1826	1972	2118	2264	2410	2556	2702	

This map is published by the World Bank Group, funded by ESMAP, and prepared by Solargis. For more information and terms of use, please visit http://globalsolaratlas.info.

Figure 3.23
Global solar irradiation in kWh/m²

It should be noted that this map is meant to give you a feel for solar irradiation. For some small systems, it may be more than enough information. However, if you are designing a small system where the amount of electrical energy delivered is critical (e.g., a solar powered vaccine refrigerator), the best data that is easiest to use to design a system is available in table and spreadsheet form (or as part of a simulation program). Generally, for non-tracking arrays (which make the best economic sense now that solar module costs have dropped so radically), you want the global insolation on a surface tilted at the latitude facing south for the northern hemisphere and facing north for the southern hemisphere. To get a map for your region, see the Global Solar Atlas.[110]

3.7. Storage and days of autonomy

If you are planning an off-grid, standalone PV system with batteries, you need to determine how important having electricity is for you. If this is a system for your home, it may demand that you spend a day without the radio if you get a string of cloudy days. However, if you need battery backup for places such as hospitals because lack of power may kill someone, you need to first determine how long you want the electricity to be provided if there is no sunlight. This is usually expressed as **"days of autonomy"** because it is based on the number of days you expect your system to provide power without receiving an input charge from the solar panels or the grid. You also need to consider the usage pattern and critical nature of your application. If it is not very important to you to have electricity every single day, then it is okay to use a small number for the days of autonomy (maybe even zero). If you are installing a system for a weekend home or a ranger station that is infrequently inhabited, you might want to consider a larger battery bank and a smaller PV array because your system will have all week (or longer) to charge and store energy. Alternatively, if you are adding a solar panel array as a supplement to a generator-based system, your battery bank can be slightly undersized since the generator can be operated if needed for recharging. This latter case would be for hybrid systems.

---

110 Global Solar Atlas https://globalsolaratlas.info/ provides information on the kW-hrs/kWp for most of the globe.

3.8. Efficiency

The efficiency of a solar cell is the percent of energy from the sun that can be converted into electricity. The efficiency of a solar cell is defined as:

Equation 3.1

$$efficiency = \frac{electrical\ energy\ out}{electrical\ energy\ in} \text{ or } \frac{electrical\ power\ out}{electrical\ power\ in}$$

The efficiency of a solar cell is reported under **Standard Test Conditions (STC)**: room temperature (25 °C), 1000 W/m^2 of solar energy with an AM1.5 spectrum. AM1.5 means an Air Mass 1.5 reference spectra ASTM G173.[111] This is shown in Figure 3.24. These distributions of power (watts per square meter per nanometer of bandwidth) as a function of wavelength provide a single common reference for evaluating spectrally selective PV materials with respect to performance measured under any type of light sources. The conditions selected for AM1.5 were considered to be a reasonable average for the 48 contiguous states of the United States of America over a period of one year. The tilt angle selected (37 degrees) is approximately the average latitude for the contiguous U.S. Depending on the time of day at your collection site, the solar energy must travel a specific distance through the atmosphere. This path length normalized to the shortest possible path length (i.e., when the sun is directly overhead) is known as the Air Mass or AM.

---

111   Reference Solar Spectral Irradiance: Air Mass 1.5 https://rredc.nrel.gov/solar//spectra/am1.5/

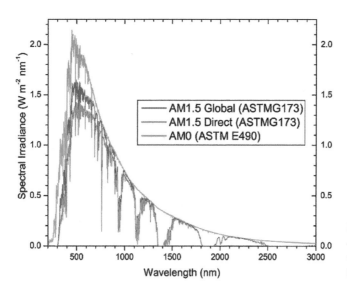

Figure 3.24
AM1.5 reference spectra used in the standard test conditions for PV.

Your PV system may never see those conditions. For example, the operating temperature of a solar cells is normally closer to 50 °C rather than 25 °C when it is outside and in sunlight. In general, the efficiency of a solar cell drops as the temperature increases. Thus, the actual energy your PV system will provide is a combination of your system design geometry with latitude and climate (solar flux and temperature). For example, a solar module with a 20% efficiency and an area of 1 m² will produce 200 W at STC, but it can produce more power on a cold, clear day when the sun is directly overhead. Likewise, that module will produce less power in cloudy conditions or when the sun is low in the sky. In central Colorado, which receives annual insolation of 5.5 kWh/m²/day (230 W/m²),[112] such a panel can be expected to produce 400 kWh of energy per year. However, in Michigan, which receives only 3.8 kWh/m²/day, annual energy yield will drop to 280 kWh for the same module.

When designing a PV system for your own use, you will want to know the rated power of the module (watt peak or Wp) more than the efficiency. Although efficiency does impact some of your costs (e.g., the higher the efficiency, the less racking you may need), remember sunlight is free, so it is not the most important factor, as we discuss in the economics sections.

112　https://www.nrel.gov/gis/solar.html

If you would like, however, you can calculate the efficiency of your panel by:

Equation 3.2

$$Efficiency\ of\ panel = \frac{peak\ power\ of\ panel}{1000\frac{W}{m^2} \times Area\ of\ panel} \times 100\%$$

The 1000 W/m^2 comes from the STC used that rated the panel's peak wattage. You can get the area of a panel by multiplying the length by the width. For example, if your panel is 1 square meter and it is rated with a peak power of 200 W, you have a 20% efficient panel.

4.   Electricity

In direct current (DC) electrical systems like a mobile phone using a battery, the electric charge (current) only flows in one direction.[113] On the other hand, in alternating current (AC) electrical systems like a blender plugged into a wall outlet, the electrical current changes direction periodically and quickly (50 or 60 times per second, depending on your country).

4.1.   DC concepts

Direct current is a single directional flow of electric charge. A battery is a good example of a DC power supply, which most people are familiar with, as they are used in everything from cars to mobile phones. You can think of the electricity stored in a battery like a water tank. You can only fill the battery (or water tank) with so much energy (or water). In Figure 4.1, the green indicator (with four bars of charges) shows that the battery is full and cannot accept any more charge. If any additional energy is directed at such a battery, this energy will be wasted. When you use energy from the battery, it is like letting water out of the tank (shown in Figure 4.1 as yellow and brown with three and two bars of charge, respectively). When the tank is empty (the battery is fully discharged), you cannot get any more water (or electricity) from your source.

---

113   Interestingly, the current flows in the opposite direction of the electrons (the direction of current is by convention the direction of flow of positive charge). If this seems a bit confusing to you – it is because it is. We can blame it all on Benjamin Franklin, the early American polymath: a leading author, printer, political theorist, politician, freemason, postmaster, scientist, inventor, humorist, civic activist, statesman, and diplomat. When he was doing his early experiments with electricity, he had a 50:50 chance of guessing the correct direction. He was a genius, but he guessed wrong anyway, so for centuries, electrical engineering has been just a little bit harder than it needs to be. So, a negative charge current leftward is an electric current rightward.

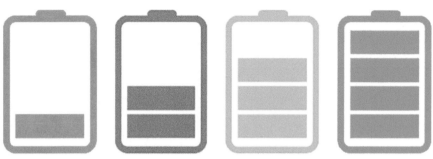

Figure 4.1
DC power storage state of charge indicators.

Solar PV also acts as a DC power supply. Direct current from PV can be used as a power supply for all kinds of electronic systems directly. However, it can also be used to charge batteries. This has the advantage that the solar electricity can be stored for later use when it is needed, even if the sun is no longer shining. The power that a PV system produces is only direct current electricity without additional electronics called inverters (which convert DC to AC).

4.2. Series and parallel

Just like normal household batteries, solar PV panels have a negative terminal (-) and a positive terminal (+). Most large PV modules now come with male and female MC4 connectors (shown in inset of Figure 4.2). The MC in MC4 stands for the manufacturer Multi-Contact, and the 4 stands for the 4 mm diameter contact pin. MC4s allow strings of solar PV panels to be easily connected by pushing the connectors from adjacent panels together by hand. There is a gratifying click when they come together. However, they require a tool (e.g., needle nosed pliers or a multitool) to disconnect them to ensure they do not accidentally disconnect when the cables are pulled either when installing or in use. Generally, the male end is positive and the female negative, but you should check the back of the module for labels on the junction box (as shown in Figure 4.2) or better yet, test with a multimeter.

The positive and negative are generally labeled on the back of the module or come with red and black color-coded wires, as shown in the inset of Figure 4.2.

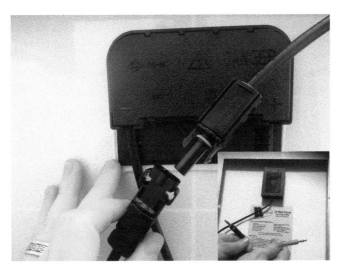

Figure 4.2
MC4 connectors on standard PV module.
Inset: Showing color coded negative and positive
terminals of small-scale PV module.

Current flows from the negative terminal through a **load** (any current-consuming device like a computer, vaccine refrigerator, light, or radio) to the positive terminal. In order to wire your solar PV panels to do something useful, you have to make a circuit. The simplest circuit is made up of a solar panel and a load, like a fan, where the current path makes a complete loop, as shown in Figure 4.3.

Panel

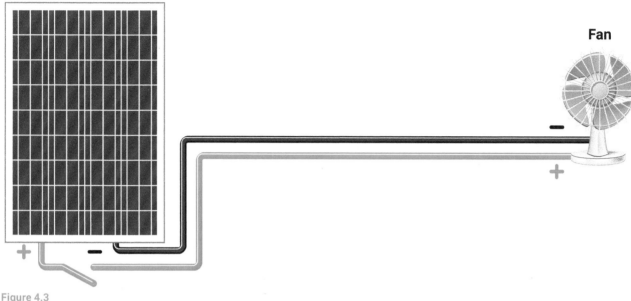

Fan

Figure 4.3
Simple PV circuit with a panel and a fan.

The current goes around the loop in only one direction: it is straight forward. However, when you start adding multiple solar panels, you have three options for how to wire them: 1) series, 2) parallel, or 3) a combination. Here we will look at the first two cases.

Series circuits have only one path for current to travel in a continuous closed loop, just as in the simple example of Figure 4.4. When circuits are arranged like this, all the current in the circuit must flow through all of the loads. In a series circuit, if you break the current flow anywhere in the loop – like if you put in a switch and flick it – the current stops everywhere. When you wire solar panels in series, you connect the positive of one solar panel to the negative of the next one and so on, as shown in Figure 4.4. When you wire solar PV panels in **series, the voltage is additive, but the current measured in Amps is constant**. As can be seen in Figure 4.4., if you wire four 12 volt and 5 amp (with a total peak power of 60 W each) solar PV modules in series, the array will be 48 volt and 5 amps (total peak power of 240 W).

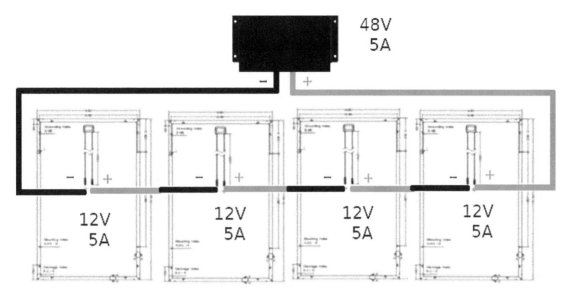

Figure 4.4
Series PV circuit.

On the other hand, when wiring solar modules in **parallel circuits**, there are multiple paths for the current to take to make a loop, as shown in Figure 4.5.

Figure 4.5
Parallel wiring of PV system.

In this type of circuit, even if you put in a switch and turn it off somewhere in the circuit, the electricity has other paths to take and can just ignore the broken path. Parallel circuits are generally used for most household electrical wiring, so when you turn off your computer, it does not turn off your lights or vice versa.

When wiring solar PV panels in parallel, the current is additive, but the voltage remains the same. So, as you can see with the same solar panels that we had in Figure 4.4, this time the voltage stays 12 V for the circuit but the current is 4 legs x 5 A/leg or 20 A. Notice that the peak power (12 V x 20 A) is still 240 W.

For the simplest circuit, a series wiring might be appropriate, and for a slightly more complex system, parallel is probably the best option. However, for more complex systems, it is also possible to use both parallel and series wiring in the same circuit. This is common for doing solar PV systems that use a string inverter. A string inverter generally has a rated voltage window that it needs from the solar PV panels in order to operate. In addition, the string inverter has a rated current to operate properly. This is because string inverters have maximum power point (MPP) trackers in them that vary current and voltage to produce the most possible (maximum) power for any given amount of sunlight. As the sunlight intensity tends to vary throughout the day, the PV array will be sending the inverter a range of both voltages and currents.

Let us assume we are designing a 480 Wp PV system with a string inverter that has a voltage window that opens at 48 V. For this system, we have access to 60 W mini-modules, which have a Voc (open circuit voltage) of 12 V. To get to the minimum voltage range, we need four modules in series.

If our system size must be 480 Wp or larger, having only four modules that are 60 W each is not enough. We need more power! We need a system with eight modules (480 W/60 W per module = 8 modules).

Thus, we will have two strings of four modules that we will wire in parallel, as shown in Figure 4.6.

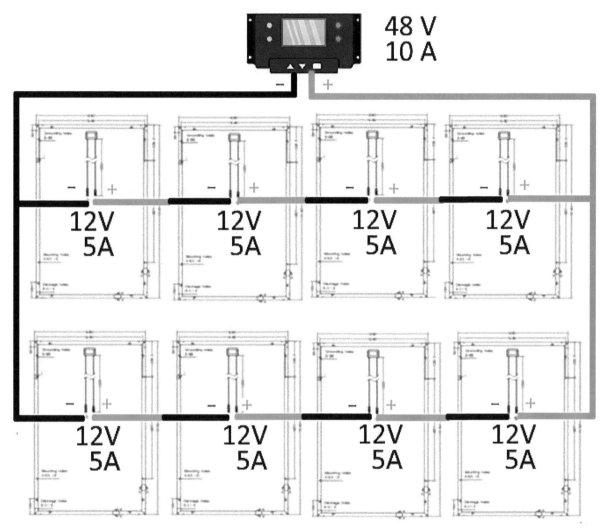

Figure 4.6
Two strings wired in series, where the strings are wired in parallel.

4.3. Power and Energy

Power and energy are often used interchangeably in the popular media, which can be really confusing. They are not the same. Officially, **electric power** is the rate per unit time, at which electrical energy is transferred by an electric circuit. That is why people often get confused – because power is equal to energy divided by time. This is exactly analogous to speed and distance. A person may have ridden their bike at 20 miles per hour (which would be analogous to a power of 20 W). That person may have ridden for 3 hours and therefore travelled 60 miles total because 20 miles/hour * 3 hours = 60 miles total. This is the same as how a 20 W device running for 3 hours would use 60 Wh of energy because 20 W * 3 hours = 60 Wh. So, power is a rate analogous to speed, and energy is a quantity analogous to distance.

Electric power is produced by solar cells. It is measured in watts. Solar cells are sold on a per unit power basis. This means you are paying a given amount of money per unit power. Wp is the power in watts for the **peak sun hours calculation we discussed before.** Power is not time sensitive: a 65 W box fan will draw 65 W of power all year (assuming it is on high). Watts are sort of small (enough power for a mobile phone or toy), so bigger devices like a car or a house may need kilowatts (kW). One kW is equal to 1000 W.

Electric energy is a particular form of energy resulting from the flow of electric charge like that which occurs when sunlight hits solar panels. Plain old energy is the ability to apply force to move an object. In the case of electrical energy, the force is applied to move charged particles (electrons). Sometimes we shorten electrical energy to **electricity** to mean the presence and flow of electric charge. Electric energy is generally measured in kilowatt-hours (kWh). One kWh in the US costs about 13 cents today.[114]

It is important to remember that power and energy are different. Your box fan may draw 65 W of power. This means that every hour that it is on, it will use 65 Wh (watt

---

114 EIA. https://www.eia.gov/electricity/monthly/epm_table_grapher.php?t=epmt_5_6_a

hours) of electric energy. It is the total energy required, not the power, which will affect the sizing of a PV array or battery bank.

An analogy often used to help understand the difference between power and energy is based on water towers.

Figure 4.7
Water tower analogy for electric energy storage.[115]

The water in the tower shown in Figure 4.7 is stored energy. When the tap is opened at the bottom, the flow of water out of the tower is power. Energy can be stored in batteries, just like the water is stored in the tower. Energy can also flow. When energy flows, it can do work like running an electric motor, operating a refrigerator, or turning on lights. The speed at which energy flows, whether it is the water flowing or the electricity flowing, is called power. The same amount of energy can be released at high power (which will occur quickly like when you press on the accelerator of an electric vehicle - put the "pedal to the metal") or at low power (which will take more time like when you are first cautiously learning to drive).

Assume that you have 1 kW of power for solar panels made up of four panels that are each 250 W. This array thus has 1000 W of total power (1 kW). Depending on where you put that 1 kW array, the energy it produces in a given day will be different. The kWh/

115 https://www.flickr.com/photos/0ccam/29763093938 Attribution-ShareAlike 2.0 Generic (CC BY-SA 2.0)

kW is the amount of energy that you can expect them to generate in a given amount of time, so you can talk about kWh/kW/year or /month or /day, etc. In Ontario Canada, over the course of a year, you can expect to get between 1,100 and 1,300 kWh/kW. So, if you found less expensive, smaller PV modules on the market and went out and bought five 200 W solar panels, you would still have 1 kW of power. If you installed them in Toronto, they would generate roughly 1,100-1,300kWh per year.

4.4. IV curves

One of the primary ways that solar cells are characterized is with an **I-V plot** where current density is graphed as a function of the voltage, as shown in Figure 4.8.

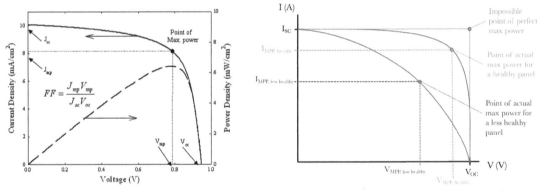

Figure 4.8

Example I-V plot and power density plot of a solar cell (left) and ideal, real, and poor example I-V curves on right.

When the voltage is zero, you have maximum current (on the far left of Figure 4.8), which is called the **short circuit current (I$_{sc}$)** and shown in Figure 4.8 as **short circuit current density (J$_{sc}$)**. If you increase voltage, the current decreases, as you can see by following the solid line to the right. If you continue to increase voltage, the current goes to zero, and this is called an open circuit, and the voltage at this point is the **open circuit voltage (V_{oc})**. In addition to the solid line following the units on the left axis, there is also a dashed line in Figure 4.8, which signifies power density and is shown on the right axis. Power can be written as:

$$\text{Power [W]} = \text{current [A]} \times \text{voltage [V]} \qquad \text{[W]} \qquad \text{(eq. 5.1)}$$

The maximum current density and voltage is at the maximum power point because after that, the power drops off, as well as the current as we move to the right of the graph. Maximum power for current is labeled as J_{mp}, and the maximum power for voltage is V_{mp}. The fill factor (FF) for a solar cell is defined as:

$$\text{FF} = (J_{max})(V_{max}) / (J_{sc})(V_{oc}) \qquad \text{(eq. 5.2)}$$

You can think of the I-V curve as a hill. A slowly sloping hill will have a low FF, which results in low power and low efficiency. A steep cliff will have higher power and higher efficiency. A 100% fill factor would have a square angle in Figure 4.8.

The IV and power-voltage curves will also be influenced by the amount of light. As the light intensity drops, so does the current and power potential of a solar cell, as shown in Figure 4.9.

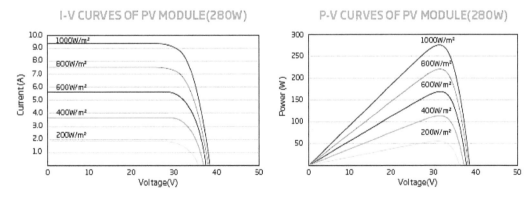

Figure 4.9
IV curves and PV curves as function of light intensity for a standard silicon solar cell.

4.5 Safety

Safety is always critical. International, national, and local organizations may prescribe their own safety measures. It is critical that you properly label and follow relevant guidelines. To give a specific example, as of this writing, designers in the US must follow laws and codes specified in Article 690 Solar Photovoltaic Systems of the National Electrical Code when installing solar systems.[116] Article 690 dives into the details of laws and regulations pertaining to solar systems and circuits within the US. Unfortunately, the language used in code can be very cryptic and/or dense, so it is always a good idea to ask for help/advice from someone who has experience installing solar in your area. With that said, here are some basic safety guidelines that you need to be sure to follow regardless of where you are.

On a final note, working with electricity is dangerous. This book is meant to help you design, size, and build small photovoltaic systems but should not be considered a guide to working with electricity. If you are just getting started, we suggest you start with a small 12 V system to avoid some of the many dangers. This section on safety is augmented by safety notes through the sections, especially those dealing with batteries which are often the most dangerous components.

4.5.1. Labeling

The benefit of labeling is that anyone who walks up to the system will have a basic understanding of what is in front of them, regardless of their experience with solar systems. A properly labeled system greatly increases the safety of the system and is a learning opportunity for anyone who is interested.

---

116   https://www.fuelcellstore.com/blog-section/standards-and-requirements-for-solar-systems

4.5.2. Wire Safety

Properly organized wiring increases safety in construction, operation, and maintenance. Here are a few tips on how to keep your wiring safe.

* Select properly insulated wires for the voltage and environmental conditions.
    * For example, exposed wire on the rooftop needs thicker insulation than indoor wires.

* Keep the color of your wiring consistent with a coloring scheme.
    * Some colors are dictated by local code.
    * Posting a legend of your coloring scheme is suggested as well.

* Properly size your wires.
    * If your wires are undersized, they will get hot and may be a fire-hazard. Pay close attention to the wire sizing section.

See Section 5.8 for more details.

4.5.3. Equipment and System Grounding

Grounding your system typically isn't required for smaller solar systems (especially those under 50 V), but if you can, you should always ground your equipment. See Section 5.9 for more.

4.5.4. Clearances/Physical Awareness

While setting up a solar system, you may find yourself in dangerous situations. Here are some general tips to increase your safety and awareness.

* If installing panels on your roof, make sure you have some kind of fall protection.

* Any live equipment (equipment that has electricity running through it) that may need maintenance (charge controller, inverter, etc.) should have proper clearance around it in case of electric shock (NEC Article 690 specifies a space 3 ft in front of the equipment and 2.5 ft wide)

✳ Open switches to panels before working on the system.

✳ Remove any metal clothing, jewelry, accessories, etc. before working on the system.

✳ Keep any flammable material as far from the system as possible.

✳ Keep your job site organized. Ideally, all components should be clearly in your field of view and easy to access (especially components that may need maintenance).

✳ Frequently visually inspect your systems after installation for any signs of wear and tear.

✳ Use insulated tools when possible.

5. Components

Photovoltaic systems are adaptable to many needs and environments. This adaptability makes them a great technology for the myriad situations a designer may encounter. This adaptability also creates an interesting challenge in generalizing a system for learning. For the purpose of this book, our base system will be small (around 100 to 500 W), with a charge controller, battery, DC loads, and AC loads, as shown in Figure 5.1. We will take out components for a simpler and smaller system (e.g., no battery) and add components for a more complex and larger system (e.g., with a lightning arrestor).

Base PV system with a charger controller, battery, DC loads and AC loads

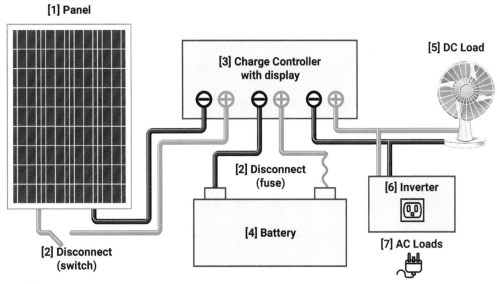

Figure 5.1

Base PV system with a [1] photovoltaic panel, [2] Disconnect (switch), [3] charge controller, [2] Disconnect (fuse or breaker), [4] battery, [5] DC loads or bank of DC breakers and plugs, [6] Inverter, and [7] AC Loads or AC subpanel. The wires and plugs are implied with the red and black lines. Not shown are other safety devices and racking. Also note that many AC and DC loads will draw too much current to be connected directly to the charge controller like this.

Each of these components are discussed in further detail in the following subsections.

[1] Panels – a collection of photovoltaic cells laminated between a clear top and an encapsulating bottom. Photovoltaic cells are thin semiconducting materials that generate power when exposed to sunlight. Shown in Figure 5.1.

[2] Disconnects – break the circuit (i.e., stop the flow of current through the circuit). Switches allow for manually breaking the circuit, whereas fuses and breakers break the circuit automatically when too much current is flowing. Shown in Figure 5.2 and Section 5.2.

[3] Charge controllers – regulate battery voltage and control the charging rate, or the state of charge, for batteries and/or loads. Often a charge controller will have an integrated meter that shows vital measurements of the system. Shown in Figure 5.1 and Section 5.3.

[4] Batteries – store the energy being generated by the panel. Shown in Figure 5.1 and Section 5.4.

[5] DC Loads – any DC device (or component or appliance) that draws power from the system. Common DC loads are anything that plugs into a vehicle cigar lighter plug and many battery powered devices. Most loads can be switched on and off. Shown in Figure 5.1 and Section 5.5.

[6] Inverters – convert the direct current (DC) generated by the panels, and/or stored in the batteries, into alternating current (AC) used by many household devices. Shown in Figure 5.1 and Section 5.6.

[7] AC Loads – any AC device (or component or appliance) that draws power from the system through the inverter. Common AC loads are anything that plugs into a wall outlet. Most loads can be switched on and off. Shown in Figure 5.1 and Section 5.7.

[8] Wires and plugs – transport the electrical current to the various components. Shown throughout Figure 5.1 and in Section 5.8.

[9] Additional safety devices – prevent shock, fire, and other risks. These devices include grounds, lightning arrestors, DC and AC breakers, and sub-panels. Shown in Figure 5.36 and Section 5.9.

[10] Racking are the mechanical fittings that physically support the panels. Shown in Figure 5.24 and in Section 5.10.

5.1. Panels

Regardless of the type, a **solar array** is made of **solar panels** (or **solar modules** so as not to be confused with solar thermal panels), and these are made up of **solar cells**.

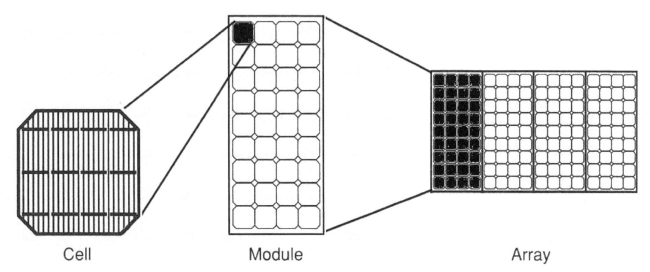

Cell Module Array

Figure 5.2
The relationship between a PV cell, module, and array.

Figure 5.2 shows drawings of the cell, the module, and the array. The cell is a single cell, and the example shown here is crystalline silicon. When you combine single cells together, you have a panel (also called a module). This is the actual electronic device you can buy off the shelf, from a supplier or the internet. The modules are wired together to get an array. The PV cell is the basic unit in a PV system. An individual PV cell typically produces between 1 and 2 W, hardly enough power for the great majority of applications. But you can increase the power by connecting cells together to form larger units called modules. Modules are normally enough to power small applications alone. If this is not enough power for you, then modules, in turn, can be connected to form even larger units known as arrays. In this way, you can build a PV system to meet almost any power need, no matter how small or large.

The most common types of solar cells are both silicon based. Crystalline silicon solar cells are circular (although they are often cut to be truncated circles or even into squares or rectangles) so they can be more closely spaced when arranging them in the module. Multi-crystalline or polycrystalline silicon solar cells are square, and in general, the multiple crystals are visible to the human eye on the surface. Thin film solar cells are extremely thin sheets of semiconductor, and they are normally integrated into the module in a way that looks like long, thin strips of material from the top. **Solar cells** generate voltage and current when exposed to sunlight. A **solar panel or module** is a collection of PV cells laminated between a clear superstrate (low iron glass, generally) and an encapsulating substrate. Most solar panels have a front of glass and a back of plastic. Some thin film cells are sandwiched between two layers of glass, while others are sandwiched between a thin sheet of metal and plastic. Solar cells can be fabricated from a number of materials, as shown in Figure 3.15.

You can learn most of what you need to know about a solar module by looking at the specification sheet from the manufacturer. Currently, Trina Solar makes the most solar modules of any company globally, so we are using them as an example here. The same information can be found in the spec sheet of any commercial PV module. Standard modules generally come in 60 PV cell module and a larger size that has 72 cells in it.[117] Both types are available in single crystalline silicon (monocrystalline) and polycrystalline. The specification sheet for the 60 cell version that has a power output of 265-285 W is shown in Figure 5.3 and Figure 5.4.

---

117 https://www.trinasolar.com/us/product

As can be seen in Figure 5.3, the electric data for a range of panel qualities (265 W to 285 W by 5 W increments) are shown for STC (1000 W/m^2, 25 °C and AM1.5), as discussed in the efficiency section 4.4. The most important value is the Wp max, which will be what you will use for calculating the performance of your system. Electricity data is also given for each Wp module under normal operating conditions (NOCT) of 800 W/m^2, ambient temperature of 20 °C, and wind speed of 1 m/s. Even though the ambient temperature is less than the cell temperature at STC, the power coming from a given cell is greatly reduced due to the lower solar flux. Care must be taken in using NOCT values for your system as well, as you may have higher operating temperatures. Realize that at STC, cells are actively cooled, and crystalline silicon-based PV are highly temperature dependent, as noted previously and shown in the temperature ratings in Figure 5.4. Note that the technical specification sheet also provides the operational temperatures that the module can work within and the fuse rating and max system voltage, which can be used to design the electrical balance of system (BOS). In addition, as shown in Figure 5.4, the mechanical data given allows for racking design and notes what type of connectors it has for electrical hookups. Similar to most modules on the market, these have a 25-year power warranty.

Electrical DATA (STC)

Peak Power Watts - P_{MAX} (Wp)*	265	270	275	280	285
Power Output Tolerance - P_{MAX} (W)			0 ~ +5		
Maximum Power Voltage - V_{MPP} (V)	30.8	30.9	31.1	31.4	31.6
Maximum Power Current - I_{MPP} (A)	8.61	8.73	8.84	8.92	9.02
Open Circuit Voltage - V_{OC} (V)	37.7	37.9	38.1	38.2	38.3
Short Circuit Current - I_{SC} (A)	9.15	9.22	9.32	9.40	9.49
Module Efficiency ηm (%)	16.2	16.5	16.8	17.1	17.4

STC: Irradiance 1000 W/m². Cell Temperature 25 °C, Air Mass AM15.

* Measuring tolerance ±3%

Electrical DATA (NOCT)

Maximum Power - P_{MAX} (Wp)	197	200	204	208	211
Maximum Power Voltage - V_{MPP} (V)	28.5	28.6	28.8	29.0	29.2
Maximum Power Current - I_{MPP} (A)	6.90	7.00	7.09	7.15	7.23
Open Circuit Voltage - V_{OC} (V)	34.9	35.1	35.3	35.4	35.5
Short Circuit Current - I_{SC} (A)	7.39	7.44	7.52	7.59	7.66

NOCT: Irradiance at 800 W/m². Ambient Temperature 20 °C. Wind Speed 1 m/s

Figure 5.3

Typical electrical data for a 24 V nominal PV module found in a specification sheet.

MECHANICAL DATA

Solar Cells	Multicrystalline 156.75 x 156.75 mm (6 inches)
Cell Orientation	60 cells (6 x 10)
Module Dimensions	1650 x 992 x 35 mm (65.0 x 39.1 x 1.38 inches)
Weight	18.6 kg (41.0 lb)
Glass	3.2 mm (0.13 inches). High Transmission. Ar Coated Tempered Glass
Back-sheet	White (PD05, PD05.08)	Black (PD05.05)
Frame	Black Anodized Aluminum Alloy (PD05.08, PD05.05)
J-Box	IP 67 or IP 68 rated
Cables	Photovoltaics Technology Cable 4.0 mm² (0.006 inches²)	1000 mm (39.4 inches)
Connector	MC4
Fire Type	Type 1 or Type 2

TEMPERATURE RATING

NOCT (Nominal Operating Cell Temperature)	44°C (±2°C)
Temperature Coefficient of P_{MAX}	-0.41%/°C
Temperature Coefficient of V_{OC}	-0.32%/°C
Temperature Coefficient of I_{SC}	0.05%/°C

MAXIMUM RATING

Operating Temperature	-40 ~ +85 °C
Maximum System Voltage (IEC)	1000 V DC
Maximum System Voltage (UL)	1000 V DC
Max Series Fuse Rating	15 A

(DO NOT connect Fuse in Combiner Box with two or more strings in parallel connection)

WARRANTY

10 years Product Workmanship Warranty
25 years Linear Power Warranty

Figure 5.4

Typical mechanical, temperature, rating, warranty, and packaging information found in a module specification sheet.

Modules or arrays by themselves do not constitute a PV system; structures that hold up the modules and electrical components make up the rest of a system and are referred to as the balance-of-system (BOS). The BOS are comprised of the additional electrical equipment and physical structures required for the PV system to operate properly. In some arrays, the modules are located directly on the roof; we will see examples of this in a moment. There is significant savings in doing this, as it avoids infrastructure such as concrete and metal poles used to support the structure of modules, which is part of the racking. In some small systems or portable systems, racking is not necessary at all. Power conditioning equipment that converts the electricity to the proper form and magnitude required by an alternating current (AC) load is also necessary. The electricity produced from solar cells is DC (and may be usable for a given application in this form), but then it can pass through an inverter in order to make it AC. When you are not on the grid, for many PV applications, you need to store the energy in some way, such as with batteries, for cloudy days or at night. Even if the array is connected to the grid, sometimes users want a battery backup for energy security reasons.

5.2. Disconnects

Disconnects is a very general term for a variety of devices that will break the connection in a circuit. Breaking the connection in a circuit, manually or automatically, is desirable for turning on and off a system, performing maintenance and repair on a system, and for overall safety to prevent shock. A switch is probably the most familiar method to manually break a circuit connection. A fuse is probably the most familiar method to automatically break a circuit connection. Disconnects come in many varieties and are rated for DC or AC and for their maximum voltage and current.

Most photovoltaic systems will require, for operation and safety, more than one type of disconnect. The most pertinent types of disconnects to photovoltaic systems are:

☀ **Switches** – manually disconnect a circuit. Shown in Figure 5.5.

☀ **DC Disconnectors** – switches usually meant for specialty applications and high voltage or high current conditions. Shown in Figure 5.6.

☀ **Fuses** – prevent component damage and fire by automatically breaking a circuit when shorted or when over current. Fuses use a small filament that permanently breaks and therefore must be replaced after each time a fuse breaks a circuit. There are types of fuses (and fuse housings) for various conditions. Shown in Figure 5.7.

☀ **Fused switches** – combine a switch with a fuse. These are usually best in high current situations but are not as convenient as a breaker.

☀ **Circuit breakers** – automatically break a circuit when the current exceeds a specific amperage and can be reset after breaking. These are mostly what you see in a household circuit breaker panel. Shown in Figure 5.8.

☀ **Ground fault circuit interrupter (GFCI)** – automatically break a circuit when detecting a difference between current coming in and current going out. These switch very fast and are specifically to protect from shock. Therefore, they are commonly installed when there is a plug near a water source. Unlike the other disconnects (which use no power), GFCIs usually consume around 1-7 watts (which can be a significant load for a small system).

Care must be taken to select the right disconnects that are available, affordable, and also match the system and safety needs. A fuse or breaker will always be necessary as close to the positive terminal of the battery as possible, since the battery is the highest source of short circuit risk. A switch or breaker is quite helpful near the panels so that you can switch off the system for maintenance or other needs. For a very small system (e.g., 50 W @ 12 V), it may not be necessary, as you can just block all the light from the panels. For a larger system (e.g. 1 kW @ 48 V), a DC disconnector and breaker may be required for safety.

Figure 5.5

DC switch (left) and AC household switches (right) where A is a 120 VAC rocker common in North America, B is a 240 VAC 4-gang rocker common in some of Africa and Oceania, C is a 230 VAC rocker common in Europe, and D is a 120 VAC toggle common in North American. Note that many countries have unique variations. appropedia.org/Six_Rivers_Charter_School_solar_cellphone_charger, appropedia.org/Light_switch_by_country

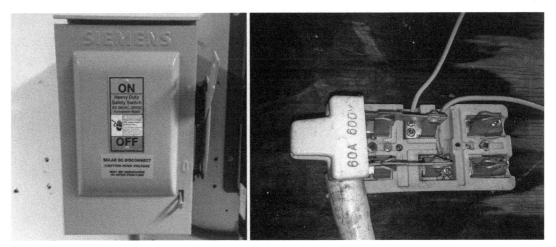

Figure 5.6

30 A Disconnect for 600 VAC or 250 VDC (left) and a 60 A for 600 VAC double pole knife switch (right). appropedia.org/CCAT_PV_system/OM appropedia.org/Ghetto2Garden_renewable_energy_2014

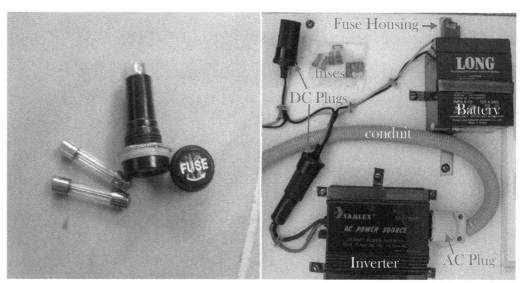

Figure 5.7

*Glass fuses (left) and blade fuses in and out of a fuse housing (right). appropedia.org/
Ghetto2Garden_solar_power appropedia.org/UTC_photovoltaic_learning_station*

Figure 5.8

*120 VAC - 10 A breaker in a breaker panel with conduit and 120 VAC plugs (left) and a 150 VDC - 63 A breaker in a makeshift
housing from a junction box (right). appropedia.org/UTC_photovoltaic_learning_station appropedia.org/Practivistas_solar_2015*

5.3. Charge controllers

A charge controller[118] regulates battery voltage and current to control the charging rate, or the state of charge, for batteries and/or loads. As discussed in Section 5.1 Panels, solar panels output varying voltage. Under full sun, the voltage from a solar panel may be quite a bit higher than the nominal voltage of the system. The need for a charge controller arises because many electronic devices are sensitive to voltage fluctuation, and overvoltage can destroy them. In addition, batteries can quickly be destroyed by overvoltage. In addition, some charge controllers use charging algorithms that can prolong the life of a battery (i.e. 3-stage charging or 'smart charging' for lead acid batteries).

Charge controllers are available in many voltages, currents, and configurations, such as: no controller, voltage regulator, basic charge controller with low voltage disconnect (LVD), pulse width modulation (PWM) controller, and maximum power point tracking (MPPT) controller.

A quick note on charge controllers: Most charge controllers have a prescribed order of connection, and disconnection, that must be followed. That order of connection is often Battery first, Panels second, Loads third. That order of disconnection is usually exactly opposite, i.e., Loads first, Panels second, Battery last.

5.3.1. No controller

In some cases, a charge controller is not necessary. The two primary cases are when (a) a DC load is not sensitive to the range of voltages from the solar panels and (b) extremely low current panels are connected to a large battery. The first case may occur with a solar fan or solar pump, for instance. The second case is called a trickle charger.

---

118   See more at https://www.appropedia.org/Photovoltaic_charge_controllers

5.3.2. Voltage regulator

A voltage regulator (Figure 5.9) is a simple charge controller that caps the voltage at a maximum voltage. A voltage regulator is usually best paired with a photovoltaic system that contains no battery, since many more complicated charge controllers need a battery. Voltage regulators are usually less expensive than more complex charge controllers. Often the power with any voltage above the cap is lost as waste heat, so there can be significant power loss if the input voltage is much higher than the output voltage. These may be as simple as a shunt or relay that redirect the energy to a big resistor.

Figure 5.9
Adjustable voltage regulator (left) that accepts inputs between 8-22 V and adjustable output between 1-15 V. USB charger (right) that accepts input voltages 9-32 V and outputs 5 V. appropedia.org/Zane_Middle_School_solar_station and appropedia.org/Six_Rivers_Charter_School_solar_cellphone_charger

5.3.3. Basic charge controller with a low voltage disconnect (LVD)

A basic charge controller (Figure 5.10) may combine voltage regulation with a low voltage disconnect (LVD). The LVD disconnects loads when the battery voltage is low, at which point the battery may be damaged. These will also often have an integrated display that shows vital measurements of the system. All charge controller systems will generally require a battery.

The maximum current is often quite limited through a low voltage disconnect. Therefore, higher power devices should still be connected directly to the battery instead of being controlled through the LVD.

Solar panels Battery and LVD loads
high power loads

Figure 5.10
Basic charge controller with a LVD. Simple LEDs indicate system status. appropedia.org/ Zane_Middle_School_solar_station

5.3.4. Pulse Width Modulation (PWM) controller

A PWM controller (Figure 5.11) rapidly connects and disconnects the panels to the batteries as a way to charge the batteries more efficiently than a voltage regulator. While PWM controllers are more expensive than basic charge controllers, they lose less power to waste than a basic charge controller or voltage regulator. These will almost always include LVDs and integrated displays as well.

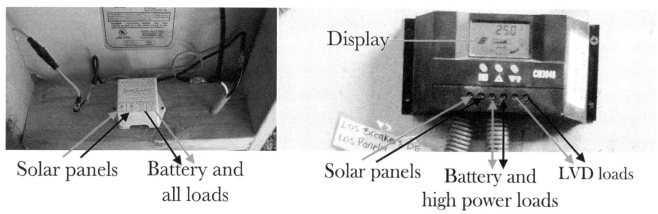

Figure 5.11

Very simple PWM controller (left) with no LVD nor display takes up to 30 V and 4.5 A in from panels and outputs 14.1 V with built in lightning protection More advanced PWM controller (right) with LVD and display takes up to 100 V and 30 A in from panels and outputs charging for a 48 V battery system with a settable float charge of 55.2 V. appropedia.org/Solar_Charged_Lawnmower and appropedia.org/Practivistas_solar_2015.

5.3.5.   Maximum Power Point Tracking (MPPT) controller

A MPPT controller (Figure 5.12) uses internal converters to charge the batteries at the required voltage while receiving power from the solar panels closest to their maximum power point seen in Figure 4.8. While MPPT controllers are more expensive than PWM charge controllers, they are even more efficient than a PWM controller. These will also almost always include an integrated display.

A significant advantage of a MPPT controller is that it allows the panel voltage to be substantially higher than the battery voltage. This may be needed due to the available panels, and it can also be advantageous because higher panel voltage will allow you to use smaller (and therefore less expensive) wire diameter from the panels.

Another advantage is that MPPT charge controllers do not need to be oversized, and in fact they can be undersized if the situation calls for it. For example, a 70 A MPPT can be connected to a 72 A solar array. The extra 2 A will be lost, reducing the system power, but nothing will break as it would with other charge controllers.

Figure 5.12
MPPT controller (vertical black rectangle in top center) takes up to 150 V and 80 A from panels and outputs charging for 12, 24, 36, 48, or 60 V battery systems. appropedia.org/CCAT_MEOW

5.3.6. Charge controller comparisons

Charge controllers are available for specific voltage systems, are constrained by a maximum current, and may have advanced features such as advanced battery charging modes. As usual, needs and budget will need to be balanced to select the best charge controller. For high voltage panels, such as those that come built for grid inter-tie, a MPPT is usually mandatory. Table 5.1 shows various aspects to consider and compare when selecting a type of charge controller.

Table 5.1
Comparisons of charge controllers.

Charge Controller Type	Cost	Efficiency	Ease of use / repair	Ease of system matching
No controller	Lowest	High	Highest	Low
Voltage regulator	Low	Low	High	Medium
Basic charge controller with LVD	Middle	Middle-Low	Low	Medium
PWM controller	Middle-High	High	Low	Medium
MPPT controller	Highest	Highest	Low	High

5.3.7. Displays

Many charge controllers come with a display (Figure 5.11 right). Displays can typically show voltage, current, power and keep track of energy.[119] In addition, a display can be added as a separate component (Figure 5.13). Displays help you track and maintain the health of a system as well as keeping track of its effectiveness.

Figure 5.13
Small DC display for voltage, current, power, and energy over time installed at Six Rivers Charter Highschool. appropedia.org/Six_Rivers_ Charter_School_solar_cellphone_charger

119 Note that most displays only track approximate energy and can drift as the battery ages.

5.4. Batteries

Batteries store energy generated from panels for later use. This can be especially important in a photovoltaic system, as loads often need energy when the sun is not out, and some days have little sun, and some loads need more power than the panels put out. A battery can allow you to store energy for when the sun is not out or even decouple generation from use. For example, a panel outputting 100 W for 5 hours generates 500 Wh. A battery could then easily power a 200 W device for a couple of hours. This means that a small panel generating for a longer time can power a larger device for a shorter time.

Currently, storage is the largest technical block to renewable energy penetration in the market. Battery technology continues to progress but significantly lags behind the progress of photovoltaic technology. As with all aspects of photovoltaic system design, many options of batteries exist to meet specific needs. These types must be chosen to fit the energy needs of the user, system voltage, lifespan, costs, and other factors.

Some key terms related to batteries in a photovoltaic system are:[120]

* **Capacity** – measure of the total energy that the battery can store. This will be listed in amp-hours (Ah) or watt-hours (Wh).[121] Batteries can be combined in series (to raise voltage) and/or parallel (to raise current) to raise the total energy storage of a system.

* **C-rate** – discharge rate of current that can be delivered by the battery. For example, A 10 Ah battery with a C-rate of: 1C can discharge comfortably at 10 A for 1 hour; 6C can discharge at 60 A for 10 minutes; or 0.5C can discharge at just 5 A for 2 hours.[122] Batteries can be arranged in parallel to provide more current.

---

120 See https://www.appropedia.org/Photovoltaic_dictionary for more terms.

121 Remember that Amps * Volts = Watts, so Amp-hours * Volts = Watt-hours.

122 The faster a specific battery discharges the lower its total available capacity is, so you may see batteries rated at different C-rates and different total capacities. https://www.appropedia.org/What_is_a_c-rate_for_batteries

* **Nominal voltage** – listed voltage of the battery that is used to match to a system.[123] For photovoltaic systems, these are typically 6 V, 12 V, 24 V, or 48 V, but other configurations are possible. Batteries can be arranged in series to raise the system voltage.

* **Charge voltage** – voltage of the batteries when at full capacity, i.e. at the end of a full charge.

* **Float voltage** – voltage to be maintained by the batteries after full charge.

* **State of Charge (SoC)** – fullness of the batteries in terms of percent total capacity.

* **Depth of Discharge (DoD)** – percent of total capacity that a battery can safely discharge to.

* **Cycle Life** – number of times that a battery can cycle to its DoD before dying.

* **Efficiency** – As with every transfer of energy, batteries incur an efficiency loss for storing and releasing their energy. Efficiency varies between battery type and state of charge.

Batteries are available in many types, chemistries, voltages, capacities, DoDs, C-rates, and cycle lives. As the technology continues to evolve, the most relevant technologies to small-scale photovoltaic systems are having no battery, a variant of a lead acid battery, or a lithium ion battery. Exciting work on storage continues to be done on ultracapacitors, saltwater batteries, sodium ion batteries, lithium sulfur batteries, block (mechanical) storage, pumped hydro storage, hydrogen storage, and solid-state batteries.

---

123 These voltages are averages, and a fully charged battery will be higher (e.g., a 12 V lead acid battery might be fully charged at 14.2 V).

Table 5.2

Table of common batteries for photovoltaics systems and some of their important properties.[124]

	Battery Type		Price	Lifespan	Depth of Discharge	Round-trip efficiency	Energy Density	Charging Rate	Self Discharge	Discharge Rate
Lead acid	Flooded	Car	Low	Low	Very Low	Medium	Low	High	High	High
	Flooded	Deep Cycle	Medium	High	Medium	Medium	Low	Medium	High	Medium
	Sealed	Absorbent Glass Mat (AGM)	High	Medium Low	Medium	Medium	Low	Low	Low	High
	Sealed	GEL	High	Medium	Medium	Medium	Low	Low	Low	High
Lithium Ion			High	Medium High	High	High	Very High	High	Low	High
Lithium Ion Phosphate			Low	High	Very High	High	High	High	Low	High

	Lifespan	Depth of Discharge	Round-trip efficiency	Energy Density	Charging Rate	Self Discharge	Discharge Rate
Low is:	200 to 900 cycles	<20%	65%-75%	30-70Wh/kg	≤0.25C	<3% / month	≤0.25C
Medium is:	900 to 1300 cycles	50%-80%	42%-85%	70-100Wh/kg	0.25V-3C	3%-6% / month	0.25V-3C
High is:	1300 to 2500 cycles	80%-100%	85%-97%	100-200Wh/kg	≥3C	>6% / month	≥3C

5.4.1. No battery

In cases where energy is needed from the solar panels only when the sun is out, no battery may be necessary. Examples of systems without a battery include some greenhouse fans, water pumping applications (Figure 5.14), and solar cell phone chargers (Figure 5.15).

---

124 https://www.appropedia.org/Table_of_battery_comparison

Figure 5.14

Students in Engineering 371 at Humboldt State University using two small solar panels in parallel to pump water. Comparing the flow rate and height of the water to the current and voltage of the panels allows for system efficiency measuring.

Figure 5.15
Solar charging kiosk at Six Rivers Charter High School. Cell phones can be charged only when the sun is out. appropedia.org/Six_Rivers_Charter_School_solar_cellphone_charger

5.4.2. Lead acid batteries

Lead acid batteries are a mature technology with well over 100 years of development and reasonably high efficiencies.[125] Conventional vehicles typically include a 12 V lead acid starter battery. Unfortunately, a car starter battery, while very available, is usually not well designed for a photovoltaic system. Car batteries (as opposed to the battery bank in an electric vehicle) have a very shallow DoD, meaning they are not meant to discharge for long. The common use case for a car battery is to discharge for only a few seconds at high current while starting the vehicle and then charge and maintain that charge while the vehicle is driving, whereas most renewable energy use cases include discharging for longer periods. Therefore, most photovoltaic systems will opt for a battery with a deeper DoD such as a golf cart battery or an even deeper DoD such as a deep cycle lead acid battery. It may be tempting to use an inexpensive car battery, but in most PV systems, you will quickly ruin them.

---

125 https://www.sciencedirect.com/science/article/pii/S2352152X17304437 85% for lead acid.

Figure 5.16
Deep cycle flooded lead acid battery (left) and sealed maintenance-free lead acid Absorbent Glass Mat (AGM) battery (right). appropedia.org/Batteries

5.4.3. Lithium batteries

Lithium batteries are an advanced battery technology with high efficiencies.[126] The two most common types are Lithium Ion and Lithium Iron Phosphate.

Lithium Ion batteries are lighter, easier to transport, and easier to recycle, and they have a deeper DoD, less loss during storage, and more cycle lives than a lead acid battery. Because of these properties, they are now common, e.g., found in power tools and laptops, the world over. They are also more expensive than lead acid batteries.

Lithium Iron Phosphate ($LiFePO_4$) is a little heavier than Lithium Ion batteries but balance that with having a higher charging/discharging rate, a longer life, and being less effected by temperature. In addition, $LiFePO_4$ is safer than Lithium Ion batteries in terms of heating up and disposal. $LiFePO_4$ is often significantly less expensive than Lithium Ion batteries.

In general, if weight does not matter, $LiFePO_4$ is a superior choice.

---

126 https://doi.org/10.1016/j.jpowsour.2019.227011 85-97% for Lithium Ion

5.4.4. Dangers

Batteries are the most dangerous component of a small-scale photovoltaic system. The most critical dangers arise from high short circuit currents, chemical spills, and off-gassing. A short circuit arises from accidentally connecting the two terminals of a battery, e.g. by dropping a metal tool across them or crossing the negative and positive wires. A short circuit will release the stored energy of the battery incredibly quickly, resulting in burns or fire. Lead acid batteries contain acids that for flooded type batteries can be spilled if tilted, over heated, or over charged. Lead acid batteries also off-gas hydrogen, which is corrosive and explosive. Flooded deep cycle lead acid batteries off-gas under most conditions and require significant venting. The sealed lead acid batteries, AGM and Gel, only off-gas a little and only when over charged, requiring them to not be installed in a sealed compartment (but the air changes for a normal human habitat are usually sufficient).

Keeping the voltage of the system low reduces the risk of shock or electrocution. For example, a 12 or 24 VDC system is usually safe to touch (unless very wet, touching the ground, and cutting open your skin to insert the wires), and systems under 50 VDC often do not require grounding. That said, it is always better to prevent contact even in these lower voltage systems that this book focuses on. In addition, make sure to check your local code for voltage guidelines.

Charge controllers protect from over charging a battery. Fuses and breakers help prevent shorts and over current. Battery terminal covers help prevent accidentally shorting a battery across the terminals. A battery box (Figure 5.17) helps prevent accidents directly related to contact with the battery. When building a battery box for a lead acid battery, remember that they off-gas and require air changes through venting.

Figure 5.17
A homemade battery box with venting and conduit.
www.appropedia.org/CCAT_greenshed_solar_lighting

5.5. DC Loads

DC loads are any DC device (or component or appliance) that draws power from the system. Common DC loads are anything that plugs into a vehicle cigar lighter plug, a USB port, and all battery-powered devices. That said, most battery powered devices come with a wall charger (see Figure 5.18) that converts AC to DC. Many DC devices are 12 V (easy to plan around), but many with wall chargers come in a wide variety of DC voltages (3 V, 5 V, 6 V, 9 V, 12 V, 18 V, etc.), and the wall charger takes care of the conversion, making it often easiest to use those as AC loads.

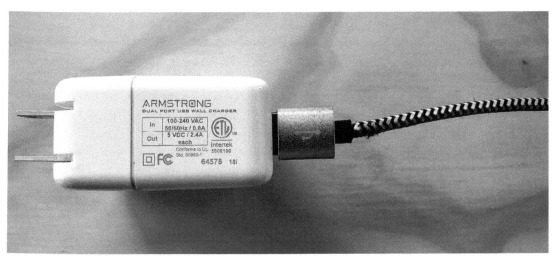

Figure 5.18
A wall charger that takes an AC input with a variable range of 100-240 V at 0.8 A and has two 5 V DC outputs at 2.4 A each.

There are many appliances available directly in DC, especially those designed for off-grid renewable systems, recreational vehicles (RVs), and, more recently, tiny homes. Most anything that plugs into a cigar lighter plug is 12 VDC, and most anything that plugs into a USB is 5 VDC.

The benefits of designing an all DC system are that you do not need to purchase an inverter and that you do not lose efficiency by inverting to AC. That said, it can be expensive (in energy and money) to need to switch between a variety of different DC voltages. Sometimes you can design a small system that is all 12 VDC or 12 VDC and 5 VDC, in which case you can save yourself from needing an inverter.

The benefits of designing an AC system is that it is easier to power your typical loads, and the higher voltage AC (e.g. 120 VAC) can use thinner, and therefore less expensive, wire to transmit longer distances (e.g., across a yard or house) than lower voltage DC (e.g. 12 VDC).

5.6. Inverters

Inverters convert the direct current (DC) provided by the solar panels into the alternating current (AC) that most household devices rely on. Ironically, many household devices actually run on DC (e.g., computers and chargers), but AC is easier to transform the voltage level and transport long distances.

As with the other aspects of photovoltaic design, many options of inverters exist to meet specific needs. These types must be chosen to fit the power needs of the user, system voltage, lifespan, costs, and other factors.

Inverters need to be matched to the system input and output voltage and frequency, as well as the maximum current needs. In addition, inverters come in a variety of waveforms. The pertinent terms related to inverters are:

✳ **Input voltage** – the nominal photovoltaic system voltage. With a battery system, simply match this to the nominal battery voltage. With a system with no batteries, you may need a voltage regulator to match the input voltage or to select an inverter with a range that matches the panels. Typical input voltages are 12 V, 24 V, 48 V, and greater than 100 V for grid intertie systems.

✳ **Output voltage** – the voltage you will be using from the system. Typically, the output voltage is based on the country you are in, e.g. 120 V in North America or 230 V in much of Africa, Asia, and Europe.

✳ **Frequency** – the frequency of the output AC voltage from the system. The frequency is based on the country you are in, e.g. 60 Hz in much of the Americas or 50 Hz in much of Africa, Asia, and Europe.

✳ **Peak (or surge) power** – the maximum power that can be drawn for a short burst of time (such as the seconds it takes to kick on a compressor in a refrigerator).

✳ **Continuous power** – the maximum power that can be drawn for sustained time periods.

* **Waveform** – the shape of the AC wave generated by the inverter. How far the waveform is from a perfect sinewave is measured as the Total Harmonic Distortion (THD) on a scale of 0 to 100%. A pure sinewave inverter might have a low THD of around 3%.

* **Efficiency** – the energy loss associated with converting from DC to AC. Efficiency in the high 90s (i.e. >90%) is common. Note that efficiency varies with load, and the average efficiency may be much lower than the maximum efficiency for a specific use.

* **Charging** – some inverters also include advanced battery charging features.

* **Alarm** – the low and/or high voltage at which the inverter will sound an alarm.

* **Cut-off** – the low and/or high voltage at which the inverter will stop working.

* **No Load Current Draw** – the current drawn by the inverter being on but not powering anything. This can often result in a power draw of 1 – 15 W (in standby) just to keep the inverter on without any load.

* **Phase** – Inverters generally come as single-phase, which is the same way it comes to most of the plugs in a typical grid connected home. Large inverters, outside the scope of this book, may come as three-phase.

* **Grid tie** – inverters designed to feed into the grid.

* **Fused, switched** – inverters that contain circuit breaking technology such as fuses or switches.

Inverters are available in many technologies, voltages, powers, waveforms, and more. The most common types of inverters for the scope of this book are no inverter, modified sine wave, true sine wave, and grid tied.

5.6.1. No inverter

Systems can be designed to be complete DC and therefore save on the costs and efficiency losses of an inverter. Keep in mind that inverters add significant adaptability and accessibility. When keeping a system DC, care must be taken to use only DC devices that match the system voltage. Alternatively, DC-DC converters can be used to match varying device voltages such as 5 V for USB and 12 V for cigar lighter plug in devices.

5.6.2. Modified sine wave

A relatively inexpensive method for inverting is to create and smooth a square wave, which is referred to as a modified sine wave (Figure 5.19). Different devices have different sensitivities to waveforms. Most household devices will function with a modified sine wave but may incur some extra efficiency loss, such as with pumps, chargers, and fans. Some devices, such as fluorescent lights, might create a slight buzzing noise, and devices that contain an internal timer based on electric frequency, such as timers and clocks, may not function correctly.

In addition, actual square wave inverters, as opposed to smooth square wave inverters, still exist but are now quite rare unless DIY. Square wave inverters are quite inexpensive but incur significant efficiency loss and are incompatible with many devices.

5.6.3. True (or pure) sine wave

A more expensive method for inverting is to create a sine wave, which is a referred to as a true or pure sine wave inverter. A true sine wave is what comes from the electric grid, so compatibility and efficiency is highest with a true sine wave inverter.

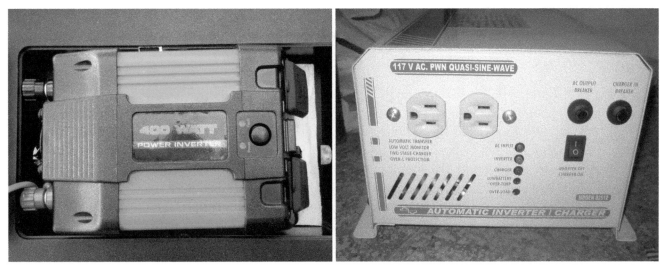

Figure 5.19
Small 400 W car style modified sine wave inverter (left) and a 1.2 kW modified sine wave inverter with battery charging capacity (right).
appropedia.org/Treadmill-a-volt_treadmill_powered_charging and appropedia.org/La_Yuca_renewable_energy_2014

5.6.4. Grid tied

Grid tied inverters are made to be connected to the mains (the main electrical power grid). These are typically household scale devices that can power the house and/or feed back into the grid. The waveform is a pure sine wave, and the frequency and voltage are designed to match the mains. Grid tie devices are typically string inverters, which connect to a string of solar panels, and microinverters that connect to each panel. Microinverters have the advantage of being more adaptable and easier to expand over time.

Some inverters also have advanced battery charging capabilities. For grid tie, this means if the mains go down, you may still have battery backup. In addition, this can couple with time-of-use electricity billing to lower your bill.

5.7.   AC Loads

Common AC loads are anything powered by a wall outlet or a circuit in a home. In a photovoltaic system, AC loads are any AC device (or component or appliance) that draws power from the system through the inverter. Different countries have different standards, but typically they are between 100 and 240 V. For example, in the United States and Mexico, the standard is 120 VAC, whereas in India and the UK the standard is 230 VAC. In addition, the AC grid is also rated in Hz, almost always either 50 or 60 Hz. The hertz (Hz) is just the frequency at which the grid voltage alternates (the A in AC). Many devices can handle either, and it is best to just stick with whatever is in your country.

It is important to keep in mind that some large AC loads, such as motors and compressors, can have a quick current surge when turned on. See the benefits of DC vs AC systems in Section 5.5.

5.8.   Wires and plugs

Wires and plugs connect all the components of a photovoltaic system into a complete circuit, i.e. delivering the current to each component. Many wires and plugs will be familiar to most readers and come in different materials, styles, and maximum voltages and currents.

The higher the max current, the larger and more robust wires and plugs you will need. You should also take special care to use plugs that are designated for the correct current and voltage so that a user does not connect an incompatible device, resulting in possible physical damage and bodily harm.

Wires come in various diameters referred to as gauge (Section 7.6 Sizing Wires and Plugs). The more current and the longer the travel length, the larger the diameter (referred to as lower gauge) you will require. Wires relevant to small-scale PV systems also come as solid, stranded, and braided, each with its own pros and cons. The terms most relevant to wires in photovoltaic systems are:

* **Solid wire** – usually less expensive, stronger (in rigid conditions), and more resistant to environmental corrosion. These are also more convenient for micro-electronics, as they will hold their form when manipulated.

* **Stranded wire** – a number of wires bundled together and usually slightly more expensive, flexible, and conductive.

* **Wire material** – the metal that the conductor is made from. The two most common are aluminum and copper. Aluminum is less expensive, less conductive, and more prone to breaking from bending. Copper is more expensive, more conductive, and more flexible.

* **Insulation** – the material surrounding the conductor that prevents accidental connection, UV degradation, heat, and moisture.

* **Color** – codified to indicate the use of the wire.

For higher current and higher voltage systems, the most current and relevant electric code should be followed (e.g., the NEC in the US, Mexico, Costa Rica, Venezuela, and Colombia). Typically, systems under 50 V do not need to be grounded,[127, 128] and systems under 30 V have even fewer restrictions.[129]

Plug type conventions should also be followed so that no one tries plugging in the wrong device. For example, do not use the standard three prong wall outlet for a DC plug.

---

127 2017 NFPA 70 NEC https://www.nfpa.org/codes-and-standards/all-codes-and-standards/list-of-codes-and-standards/detail?code=70 250.20 (A) lists grounding conditions for systems less than 50 V (mostly relates to transformers). 250.20 (B) lists grounding conditions for 50 V to 1000 V (more stringent than <50)

128 https://www.ecmag.com/section/codes-standards/know-rules breaks down 2017 NEC rules for grounding low voltage systems

129 https://www.ecmweb.com/code-basics/solar-photovoltaic-systems-part-2

5.8.1. Insulation

Wires carrying higher currents must be larger, and wires carrying higher voltages must have stronger insulation. In addition, environmental conditions and use must be considered. Different countries have different wire type codes. For example, in the U.S., THWN stands for Thermoplastic Heat and Water-resistant Nylon-coated. For extra protection, conduit can be used to house the wire.

Wires that will be on a roof should be insulated for exposure to moisture, UV, and heat. Many panels come with PV wire or USE-2 wire, which are both sufficient for roof installation.[130, 131]

5.8.2. Color

Color indicates the use of the wire. It is important to follow local norms to increase the ease of maintenance and repair and to decrease the chance of accidents. In addition to color, labels greatly add to the clarity, safety, and longevity of a system.

DC color codes for small scale systems are: equipment ground is green or bare, negative or conductor ground is black, and positive is red (or a different color if needed). AC color codes vary widely by country. For example, in the U.S., hot (aka the current source) is red or black (or blue for a 3rd phase that we do not cover in this book), neutral (the return path for the current source) is white, and ground is bare or green.

---

130 "The 2017 NEC Article 690 Part IV Wiring Methods permits various wiring methods in photovoltaic systems. For single conductors, UL Listed USE-2 (Underground Service Entrance) and PV wire types are permitted in exposed outdoor locations in PV source circuits within the PV array. PV wire is further permitted to be installed in trays for outdoor PV source circuits and PV output circuits without needing to be rated for tray use. Restrictions do apply if the PV source and output circuits are operating over 30 volts in accessible locations. In these cases, Type MC or suitable conductors installed in raceways are required" https://www.anixter.com/en_us/resources/literature/wire-wisdom/pv-wire.html

131 "The insulation covering wire can protect the cable from heat, moisture, ultraviolet light or chemicals.
- THHN is commonly used in dry, indoor locations.
- THW, THWN and TW can be used indoors or for wet outdoor applications in conduit.
- UF and USE are good for moist or underground applications.
- PV Wire, USE-2 and RHW-2 cables can be used in outdoor and wet conditions where their outer cabling is UV and moisture resistant. They must be sunlight resistant." https://www.civicsolar.com/support/installer/articles/solar-wire-types-solar-pv-installations

Care should be taken to connect or solder wires and plugs correctly. Junction boxes and housings will also increase the safety. An old trick to increase waterproofing is to allow the wire or conduit to bend down and back up so that water drips off where you design the bend instead of following the wire into a housing.

Figure 5.20
Three 160 W flexible monocrystalline panels with MC4, and MC4 Branch, Connectors. Dusty Build by Tressie Word

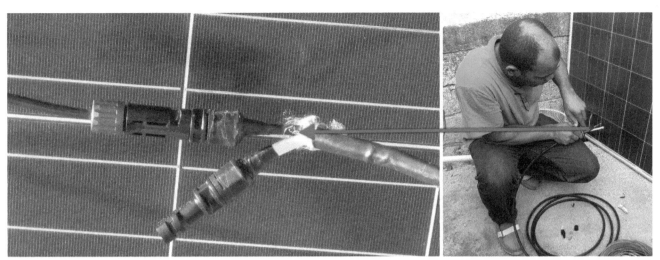

Figure 5.21

MC4 solar panel connectors (left) and wires being stripped (right). appropedia.org/Practivistas_solar_2015

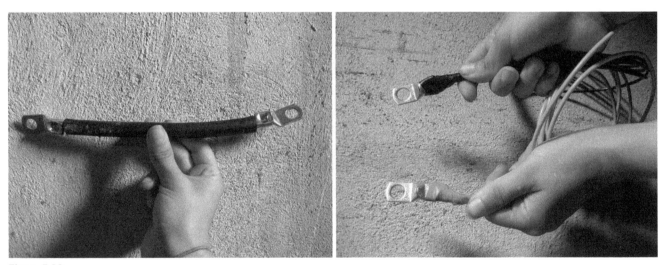

Figure 5.22

Large diameter wire (jumpers) with terminals for connecting batteries together (left) and connecting the batteries to the rest of the system (right). appropedia.org/Practivistas_solar_2015

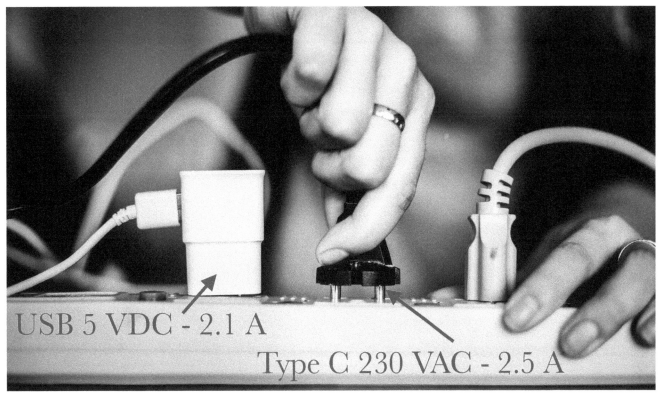

Figure 5.23
USB 5 V DC - 2.1 A max plug common worldwide (left) and a Type C 230 VAC - 2.5 A max common in Asia, Europe, and South America (right)

5.9. Additional safety devices

Safety is critical in all designs, especially those with electricity. The larger the system, the higher the voltage, the more batteries, the more general the users, the more public the access - all reasons to increase the safety.

Additional safety devices include meters, signage, alarms, physical grounds, conduit, junction boxes, and lightning arrestors. We suggest you consider these when designing your system.

5.10. Racking

Racking (or mounting) are the mechanical components that hold a PV panel in place.

Here we will look at the basics of PV racking as well as two methods anyone can use to fabricate both a simple, low-cost ground mounted (or flat roof) PV system and a building-integrated PV system for their own rooftops.

5.10.1. Basics of Racking

In the not so distant past, racking played almost no role on the impact of the costs of a PV system. However, as PV module costs ($/W) have been driven into the floor, the economic role of other BOS components have gained prominence. In fact, now racking can even dominate the economics of a PV system.[132] Due in large part to this lack of attention and the relative unimportance of the costs of PV racking, it has been marginalized historically, and there has been little progress on reducing the materials and costs associated with PV racking. This presents an enormous opportunity because most communities have both the skills and materials to fabricate a basic rack, which can radically reduce the cost of a small PV system.

---

132 Barbose, G., 2014. *Tracking the Sun VI: An Historical Summary of the Installed Price of Photovoltaics in the United States from 1998 to 2012.* L. B. N. Laboratory.

5.10.2.   Installation angle

If you are installing the modules for only a season, you can use this simple rule of thumb: Take your latitude and add 15 degrees for the winter, or subtract 15 degrees for the summer. For example, if your latitude is 45 degrees, the angle you want to tilt your panels at in the winter is: 45 + 15 = 60 degrees. In the summer, it would be: 45 - 15 = 30 degrees. You do not necessarily need to change the angle throughout the year, but you can get an exact optimum tilt angle using a free simulation program like System Advisor Model (SAM).[133] SAM can help you economically optimize large projects but can be overkill for most small projects.

5.10.3.   Types of DIY Racking

You can either make your own racking or buy a ready-made system. If you do the latter, you will pay for it – normally now more than the module for a small-scale system. If you make it yourself, you will save some significant capital costs, but that depends on what option of **DIY** racking you choose. Below are some **DIY** options in increasing levels of sophistication:

✳ For the simplest system, particularly if it is temporary, you can just prop the modules up against something such as a log, pallets (as shown in Figure 5.24), the side of a building, or a large rock, etc.

✳ A slightly more sophisticated **DIY** portable system is to make it from PVC pipe and hold the module to it with zip ties.[134] The key is to point the PV panel at the sun.

✳ For the next step up in **DIY** sophistication, one can make a single panel mounting system from 2x4s or scrap wood[135] like that from a pallet (see Figure 5.24).[136] Many people have made wooden PV racks[137] that work fine, although they may not last as long as those commercial systems that are generally made from metal even when sold

---

133   https://sam.nrel.gov/opensource and https://sam.nrel.gov/download

134   https://gnomadhome.com/pvc-solar-mount/

135   https://waldenlabs.com/diy-off-grid-solar-system/

136   https://www.youtube.com/watch?v=TQQGVN9JKdQ

137   https://www.builditsolar.com/Projects/PV/EnphasePV/Mounts.htm

as DIY kits.[138] Wooden racking systems can be sophisticated, such as the adjustable mounting system shown in Figure 5.26.[139]

✳ The next level of sophistication uses some wood and some metal, as in unistrut[140] or simple extrusions.[141] You can also make a super low-cost floating PV system with marine epoxy and foam (see Figure 5.25).[142]

✳ Another level of sophistication uses 3D printed parts covered in the next section.

✳ For a more detailed treatment of PV racking see https://www.appropedia.org/Open_source_DIY_PV_racking

Figure 5.24
Screen capture of DIY video for solar panel mounting on pallets by David Poz.

138 https://www.doityourselfsolarracking.com/

139 https://www.instructables.com/id/Tiltable-Solar-Panel-Mount/

140 http://www.tincancabin.com/2013/08/more-solar/#more-1797

141 http://www.ottawavalleypv.ca/low_cost_ground_mounts.html

142 Mayville, P., Patil, N.V. and Pearce, J.M., 2020. Distributed manufacturing of after market flexible floating photovoltaic modules. *Sustainable Energy Technologies and Assessments, 42*, p.100830. https://www.researchgate.net/publication/346541241_Distributed_manufacturing_of_after_market_flexible_floating_photovoltaic_modules

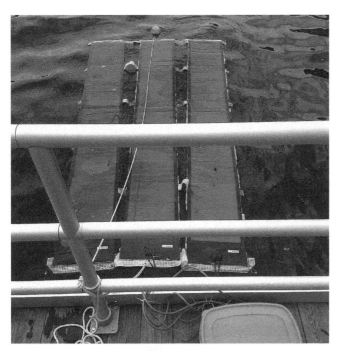

Figure 5.25
Flexible PV modules converted to FPV.

Figure 5.26
Adjustable wooden solar panel mount. CC-BY-NC-SA by Instructable user Born_to_build.

5.10.4.   3D Printed Racking

The technological evolution of the 3-D printer, widespread internet access, and inexpensive computing has made a new means of open design capable of accelerating self-directed sustainable development.[143] A 3-D printer is a machine that builds up 3-D objects one layer at a time. 3-D printers are pretty interesting as they enable the use of designs in the public domain to fabricate open source appropriate technology (OSAT),[144] which are easily and economically made from readily available resources by local communities to meet their needs. The RepRap[145] project (self-Replicating Rapid Prototyper) pushed the price down to a few hundred dollars.[146] There have also been several types of PV powered RepRaps specifically for humanitarian crisis situations.[147, 148, 149, 150] One way to decrease the racking costs for a PV system is to use a RepRap 3-D printer to make it for small-scale mobile PV arrays.[151] A study[152] evaluated the technical and economic viability of distributed manufacturing of PV racking in the developing

143 J. M Pearce, C. Morris Blair, K. J. Laciak, R. Andrews, A. Nosrat and I. Zelenika-Zovko, "3-D Printing of Open Source Appropriate Technologies for Self-Directed Sustainable Development", *Journal of Sustainable Development* 3(4), pp. 17-29 (2010). https://doi.org/10.5539/jsd.v3n4p17

144 Joshua M. Pearce, "The Case for Open Source Appropriate Technology", *Environment, Development and Sustainability, 14, pp. 425-431 (2012). https://link.springer.com/article/10.1007/s10668-012-9337-9*

145 https://reprap.org/wiki/RepRap

146 Wittbrodt, B. T., Glover, A. G., Laureto, J., Anzalone, G. C., Oppliger, D., Irwin, J. L. & Pearce, J. M. (2013). "Life cycle economic analysis of distributed manufacturing with open-source 3-D printers", *Mechatronics*, 23, 713-726. https://www.academia.edu/4067796/Life-Cycle_Economic_Analysis_of_Distributed_Manufacturing_with_Open-Source_3-D_Printers

147 Debbie L. King, Adegboyega Babasola, Joseph Rozario, and Joshua M. Pearce, "Mobile Open-Source Solar-Powered 3-D Printers for Distributed Manufacturing in Off-Grid Communities," *Challenges in Sustainability* 2(1), 18-27 (2014). http://dx.doi.org/10.12924/cis2014.02010018

148 Benjamin L. Savonen, Tobias J. Mahan, Maxwell W. Curtis, Jared W. Schreier, John K. Gershenson and Joshua M. Pearce. Development of a Resilient 3-D Printer for Humanitarian Crisis Response. *Technologies 2018, 6(1), 30;* https://doi.org/10.3390/technologies6010030

149 Khan, K.Y., Gauchia, L., Pearce, J.M., 2018. Self-sufficiency of 3-D printers: utilizing stand-alone solar photovoltaic power systems. *Renewables: Wind, Water, and Solar 5:5.* https://doi.org/10.1186/s40807-018-0051-6

150 Jephias Gwamuri, Dhiogo Franco, Khalid Y. Khan, Lucia Gauchia and Joshua M. Pearce. High-Efficiency Solar-Powered 3-D Printers for Sustainable Development. *Machines 2016, 4(1), 3;* https://doi.org/10.3390/machines4010003

151 Wittbrodt, B., Laureto, J., Tymrak, B. and Pearce, J.M., 2015. Distributed manufacturing with 3-D printing: a case study of recreational vehicle solar photovoltaic mounting systems. *Journal of Frugal Innovation, 1(1), p.1.* https://jfrugal.springeropen.com/articles/10.1186/s40669-014-0001-z

152 B.T. Wittbrodt & J.M. Pearce. 3-D printing solar photovoltaic racking in developing world. *Energy for Sustainable Development 36, pp. 1-5 (2017).* https://www.researchgate.net/publication/309883530_3-D_printing_solar_photovoltaic_racking_in_developing_world

world as well as flat roof buildings[153] using entry-level RepRap 3-D printers and found some promising results.

An example 3-D printed racking system is the X-wire system works which uses the existing aluminum frame of the standard PV module as structure and keeps the module in place with end brackets pulled against the frame with steel cable. A 1 kW PV array consisting of four 250 W PV modules, shown in Figure 5.27, using the X-wire racking system.

Figure 5.27

Assembled 1 kW PV array with X-wire system. The 3-D printed components shown in grey are at the corners. The wiring forms an x-pattern beneath the modules.

When compared to a commercial racking system, the X-wire system is significantly less expensive, with a savings of 83% (with commercial 3-D printing filament) to 92% (with recycled waste plastic). When fabricating racking yourself, you can also avoid import duties, which can be important, depending on your country. With the X-wire system, the largest individual cost is the printed plastic, with 1.5 kg/kW used at $33/kg. Using a new technology, the recyclebot,[154, 155] which converts waste plastic to 3-D printer

153 B.T. Wittbrodt & J.M. Pearce. Total U.S. cost evaluation of low-weight tension-based photovoltaic flat-roof mounted racking. *Solar Energy 117* (2015), 89–98. https://www.academia.edu/29779538/3-D_Printing_Solar_Photovoltaic_Racking_in_Developing_World

154 http://www.appropedia.org/Recyclebot

155 Baechler, C., Devuono, M. & Pearce, J. M. (2013). "Distributed recycling of waste polymer into RepRap feedstock", *Rapid Prototyping Journal, 19,* 118-125. https://www.researchgate.net/publication/235703067_Distributed_Recycling_of_Waste_Polymer_into_RepRap_Feedstock

feedstock, the cost is cut further to about ten cents per kg. Recyclebot extruded filament is particularly applicable in developing regions, as there have already been efforts to create ethical filament standards,[156] which would allow waste pickers (those who make a living digging through landfills for recyclable materials) to lift themselves out of poverty by capturing a larger share of the value from recycling plastics into 3-D printer filament.

5.10.5. After market Building Integrated PV (BIPV) Racking

A typical residential roof rack comes in many different styles, depending on roof type (shingles, metal, tile, etc.), but they all have similar traits. Currently, all of the rooftop PV mounting systems used for conventional modules have some sort of standoff, which will increase the distance between the roof and the PV modules from inches to feet, depending on roof slope. The standoff has been historically included in order to dissipate the thermal energy in the modules in an effort to minimize efficiency loss from higher operating temperatures for the PV. There are many racking options for different appearances and angles for optimal settings, and their costs are now relatively high compared to reduced module costs. For example, the Rooftrac racking system costs $0.12/W for a residential application.[157] Such inflated racking costs can make systems uneconomic.

In addition, such systems always demand the full cost of a standard roof underneath them, as the PV do not act as a roof. This is a big waste! The PV industry has been trying to fix this problem, with a growing number of building integrated photovoltaic (BIPV) systems, with products such as PV roof slate and PV roof shingles. In these cases, the PV becomes your roof – so you no longer need to pay for the roof. These BIPV products generally cost far more than conventional PV systems primarily because of their relatively small scale that cannot take advantage of the economies of scale observed in the standard module market. In order to overcome these challenges, a study investigated a novel modification of conventional PV modules to allow them to act as BIPV roofing

156 Feeley, S. R., Wijnen, B., & Pearce, J. M. (2014). Evaluation of Potential Fair Trade Standards for an Ethical 3-D Printing Filament. *Journal of Sustainable Development*, 7(5), 1-12. https://www.researchgate.net/publication/285936892_Evaluation_of_Potential_Fair_Trade_Standards_for_an_Ethical_3-D_Printing_Filament

157 "Rooftrac- Residential Roof Solar Mounting System, Solar Racking". ProSolar- Solar Mounting Systems - Oxnard, CA. N.p., 2017.

slates, (full details can be found in the open access study).[158] The designs are available as fully open source hardware at the Open Science Framework.[159]

5.11. Monitoring

Monitoring allows you to quickly check and track system health. Many charge controllers and inverters have built in monitors, but you may find that you want to track specific loads or charging and discharging over time. Some charge controllers and inverters can also upload their data to the web.[160]

5.12. Systems

There are basically three types of photovoltaic systems, each of which have a specific sizing requirement. The standalone systems use photovoltaic technology only and are not connected to a utility grid and therefore usually have some form of backup, such as a battery. The second type of system is the hybrid system. Hybrid systems are made up of PV and other forms of electricity production, such as wind or diesel generation. Finally, grid-tied systems do not necessarily need backup at all and are tied directly to the utility grid, which 'stores' the energy. For example, if you have a PV system on your house that is connected to the electric grid, and it produces more energy than needed, the extra will

158 Pearce, J., et all 2017. Design of post-consumer modification of standard solar modules to form large-area building-integrated photovoltaic roof slates. *Designs*, *1*(2), p.9. https://doi.org/10.3390/designs1020009

159 CAD assembly for module BIPV retrofit. Open Science Framework. https://osf.io/pz3uv/

160 See examples at https://www.appropedia.org/Photovoltaic_monitoring.

go back onto the grid. So, during cloudy days or at night, you can pull the energy from the grid into your home. You can also have a grid connected system with batteries, but will need special permission and equipment in order to run it as an 'island' in instances where the grid power goes out.

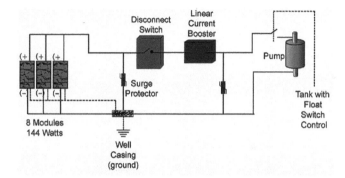

Figure 5.28
Standalone PV system design for water pumping with no storage.

Figure 5.29
Solar water pumping in India (NREL).

Figure 5.28 shows an example of a standalone PV system similar to the one used to pump water in India, as shown in Figure 5.29. The basic system has the PV array, which produces electricity that then travels through a disconnect switch and linear current booster to provide power to the water pump. This system is common in agricultural applications throughout the world, such as the farm in Thailand shown in Figure 5.30.

It is very simple to design and has no electrical storage. It should be noted, however, that this system can be thought of as having storage in the form of pumped water, whose potential energy has been increased by the pumping process. Therefore, the system can produce pumped available water, which can be stored and used when required. This is, in many ways, a more reliable and inexpensive way than storing the electricity in batteries to be used to pump water on demand. The depth of the well and the amount of water needed on a case-by-case basis determine the size of the system because of the power requirements of the pump. Standalone systems do not always need to be for water pumping. This 20 W system in Tibet can be used to charge cell phones and laptop computers. The batteries in the devices are being charged, and charging is only possible during sunny days.

Figure 5.30
PV water pumping on large system in Thailand (NREL).

Figure 5.31
Small standalone PV system in Tibet (NREL).

The next type of system is the hybrid system. If you need electricity immediately, PV-generated electricity runs through the inverter and if needed a transformer and goes to your load. Otherwise, it is stored in a battery. Then, if it becomes dark out, you can draw the energy from the battery. If it is dark for a very long time, or if the load exceeds what is available from the PV panels, you can draw the energy from a generator. In the ideal case from an efficiency standpoint, the generator would be a co-gen (i.e., supplying useful heat and electricity). In general, such systems tend to be larger, as they are viewed as providing a critical supply of electricity (e.g., for a communication system or hospital).

Most PV systems installed today are grid-tied systems for those of us who are connected to grid system houses; if you don't live in a rural area, you can probably get your electricity from a grid. A common household system type for much of the world is a net-metered setup, as shown in Figure 5.32. Here you have solar panel array on your roof, and the sun shines on it and produces a DC voltage (like a battery). Then the DC voltage runs through the inverter and turns into AC voltage (like the electricity that comes from a wall outlet). The inverter is usually monitored in some way, and usually you'll have a breaker panel inside your home. AC voltage goes straight to the main utility breaker panel. From there, it can be used to power any electrical device in your home. In some cases, let's say, at lunch time, if you're at work, and the solar cell is producing electricity, you're going to be producing a lot of electrical energy, and then it can be fed back into the grid and 'spin your utility meter backwards' (which is a somewhat antiquated term, but a fitting visual). You then pay only for the net amount of electricity you use. This is called **net metering**. Some solar users purposely make more electricity than they use so they can receive money back from the electric companies. A two-meter system is similar to this setup, with the only difference being that it is not connected to the house's breaker and feeds all of its electricity onto the grid.

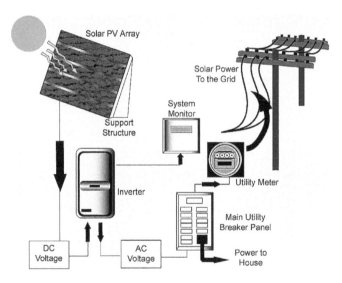

Figure 5.32
Schematic of grid tied PV system. Solar panels convert sunlight directly into electricity. The Inverter converts the solar electricity (DC) into household current (AC) that can be used to power loads in the house. The System Monitor is an easy-to-read digital meter that shows the homeowner the amount of electricity generated both cumulatively and daily. The Utility Meter tracks power usage and production, spinning forward when electricity is used from the grid, and spinning backwards, generating a credit, when the solar system creates more electricity than is used in the house.

No matter which type of solar system you are considering, you can actually design it yourself. Most of the instructions to follow pertain to designing an off-grid system with battery backup; however, many of the principles can be used for sizing an on-grid system or a simple water pumping system as well.

Photovoltaic systems can vary greatly in scale and scope. Systems under a watt can power a watch, and systems in the megawatts can power communities. This book focuses on small-scale systems, usually under 1000 W (also known as 1 kW). The scope of a system can vary from powering one simple device to many complex devices.

The system in Figure 5.33 shows a very basic system where a panel (1) directly powers a DC load (5). A disconnect (2), probably a basic switch, allows for the power to be turned off when needed. The advantages of this system are that it is inexpensive, simple, and durable. The disadvantages are that it functions only when the sun is out, and the load must be able to handle the large voltage fluctuations that the panel will provide.

Some classic examples of systems without batteries include a greenhouse fan and a water pump. A greenhouse fan serves to cool down a greenhouse during the day. The greenhouse is hotter when the sun is out in full-force, so the fan will be working more due to the same sunlight. A water pump can be set to pump up to a higher reservoir only while the sun is out. Since the higher reservoir works as storage, the pump can work when the sun is out instead of on demand.

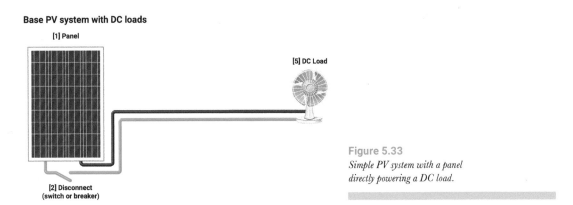

Base PV system with DC loads

[1] Panel

[5] DC Load

**[2] Disconnect
(switch or breaker)**

Figure 5.33
*Simple PV system with a panel
directly powering a DC load.*

The system in Figure 5.34 shows a basic system where the voltage from a panel (1) is regulated by the voltage regulator[161] (3) and powers a DC load (5). A disconnect (2), probably a basic switch, allows for the power to be turned off when needed. The advantages of this system are that it protects the DC load and is still relatively inexpensive,

---

161 Remember that in this book, we refer to voltage regulators as a limited type of charge controller.

simple, and durable. The disadvantages are that it functions only when the sun is out and adds the cost of a charge controller.

Some typical uses for this type of system include a USB charger for mobile phones (and other 5 V devices), powering LED lights (e.g., in a place where natural lighting is not an option, such as some offices and favelas), and powering devices that have internal batteries (e.g., a portable speaker). This type of voltage protection from the voltage regulator may also be necessary for the fan and water pump discussed previously, if that fan and water pump are sensitive to voltage.

Base PV system with a voltage regulator and DC loads

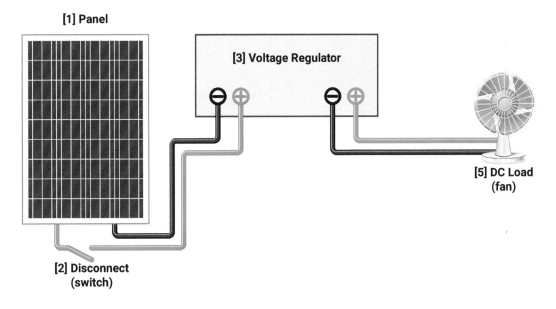

Figure 5.34

PV system with a panel and voltage regulator powering a DC load.

The system in Figure 5.35 shows a system where the voltage from a panel (1) is regulated by the charge controller (2) to charge a battery (4) and power a DC load (5). A disconnect (2) allows for the power to be turned off when needed, and a fuse (2) provides over-current protection. The advantages of this system are that it protects the DC load and stores energy for use when the sun is not available. The disadvantages are that it is more expensive and dangerous due to the addition of a battery.

This type of system can be used for all types of DC devices, such as mobile phones, LED lights, water pumps, and any device that plugs into a cigar lighter plug in a vehicle. In addition, the battery allows for powering devices that take more power than the panel can provide, albeit for less time (because the battery will be draining faster than the solar power can fill it).

Base PV system with a charge controller, battery, and DC loads

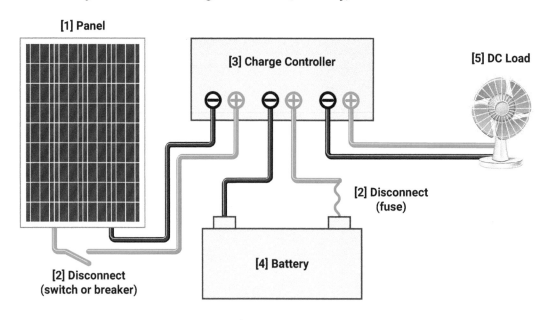

Figure 5.35
PV system with a panel and charge controller charging a battery and powering DC loads.

The system in Figure 5.36 shows a system where the voltage from a panel (1) is regulated by the charge controller (3), probably one with a display, to charge a battery (4) to power DC loads (5) and an inverter (6) to power AC loads (7). A disconnect (2) allows for the power to be turned off when needed, and a breaker (2) provides over-current protection, while a ground (8) and a lightning arrestor (8) provide extra system protection. The advantages of this system are that it protects the DC and AC loads and stores energy for use when the sun is not available. The addition of the inverter provides AC power for common household devices. The disadvantages are that the addition of a battery, inverter, and extra protection make it more expensive, and the battery makes it more dangerous.

This type of system can be used for all the typical DC devices as well as the AC devices commonly found in households. This type of system is often larger than the other types of systems and can be scaled up to household or larger sizes.

Somewhat larger PV system with a charge controller, battery, lightning arrestor, DC loads and AC loads

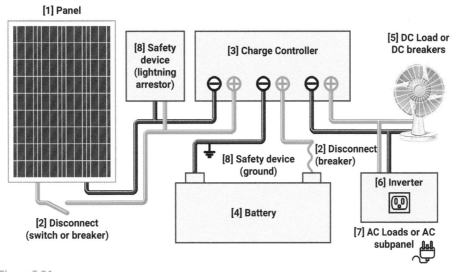

Figure 5.36
PV system with a panel and charge controller charging a battery and powering DC and AC loads.

Often the DC and AC loads are too large to control from the charge controller directly. In that case, the loads can be connected to the battery.

The system in Figure 5.37 shows a system where the voltage from a panel (1) is regulated by the charge controller (3), probably one with a display, to charge a battery (4). That battery has a disconnect for the charging from the charge controller and a separate disconnect for the loads being powered directly from the battery, i.e. the DC loads (5) and inverter (6) to power AC loads (7). A panel disconnect (2) allows for the power to be turned off when needed. A battery breaker (2) provides over-current protection. A load's fuse provides protection from shorting, while a ground (8) and a lightning arrestor (8) provide extra system protection. The advantages of this system are that it protects the DC and AC loads and stores energy for use when the sun is not available. The addition of the inverter provides AC power for common household devices. The disadvantages are that the addition of a battery, inverter, and extra protection make it more expensive, and the battery makes it more dangerous. Connecting the loads to the battery retains the advantage of regulated storage but loses the ability for the charge controller to regulate the loads and shut them off in the case of the battery discharging too low.

Larger PV system with a charge controller, fused battery, lightning arrestor, fused DC loads and larger inverter not controlled by charge controller

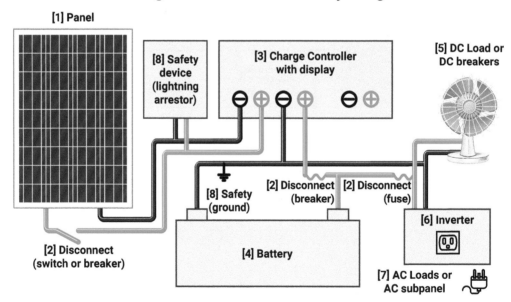

Figure 5.37

PV system with a panel and charge controller charging a battery and powering DC and AC loads directly from the battery.

PV systems can also now be used to provide heating and cooling by coupling them to heat pumps, as shown in Figure 5.38. A heat pump is a device that uses electricity to transfer heat energy. The most common heat pumps are used to keep refrigerators cold. Now, they can also be used to both heat and cool a building more efficiently because their coefficient of performance can be higher than one (e.g., they can use one unit of electrical energy to move more than one unit of heat energy). Although solar thermal systems can be effective, it is easier to scale PV-powered heat pumps, given that the systems can be installed and operated independently. PV and heat pump technologies have finally come down low enough in price that they can beat the cost of natural gas furnaces and the grid in Canada and the U.S., where the heat pump is primarily used for heating.[162] In warmer climates, the heat pump can be used for air conditioning, or if you live somewhere in the middle, it can be used for heating or cooling depending on the time of year.

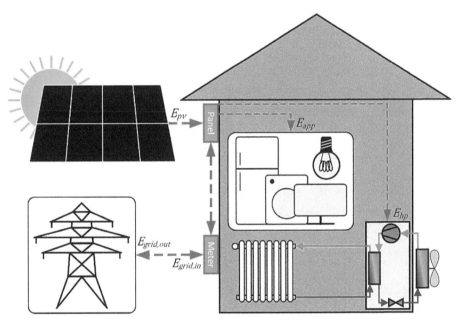

Figure 5.38

Systems diagram of building energy system with PV generation (Epv), appliance loads (Eapp), heat pump loads (Ehp), and a grid connection (Egrid). https://www.appropedia.org/File:PV_HP_diagram.png

162 Joshua M. Pearce and Nelson Sommerfeldt. 2021. "Economics of Grid-Tied Solar Photovoltaic Systems Coupled to Heat Pumps: The Case of Northern Climates of the U.S. and Canada" *Energies 14, no. 4: 834.* https://doi.org/10.3390/en14040834

6. Sizing a system with no batteries

Sizing a system is often an iterative process based on energy needs and access to resources. The first step in sizing systems should probably always be to conserve first. It is almost always less expensive to conserve than to overproduce. This is sometimes referred to as negawatts (the power you avoid).

The next thing you will want to decide is whether offgrid or not to have batteries. This section is for sizing a system with no batteries. Most systems will need a battery, but systems without batteries are generally safer, less expensive, and longer lasting. Section 5.4 will help you determine if you want batteries. If your system needs a battery, skip this section, and go to Section 7.

6.1. Simple DC load

The simple **DC** load photovoltaic system powers a load directly from solar with no regulation or storage. Not many systems will work like this, but when it works, it is usually the least expensive and easiest to use system. Some typical uses for this type of system include a **USB** charger (with built in regulation) for mobile phones (and other 5 V devices) and **LED** lights (in a place where natural lighting is not an option, such as some office buildings and favelas). As stated earlier, this type of system will provide power only when the sun is out. The device will not work when the sun isn't out, unless it has its own internal battery.

To size this system, there are two main constraints: voltage and current.

Voltage: Devices have a voltage range at which they will operate. Some devices have a very narrow range (e.g., a laptop that might charge with a DC-in of exactly 19 V), while other devices might have a large range (e.g., some low voltage LED drivers can accept 10-32 V). You must ensure that the load(s) you want to power can accept the whole range of likely voltages from the panel(s) you select. If your panels have a VOC (open circuit, aka max, voltage) of 21 V, then your device in this setup will need to be able to accept that voltage.

Current: Devices have a required current to operate well. It is usually not a problem to use a panel that can provide a little more current (I_{MPP}, the maximum power point current is the best to look at for this), as the device(s) will only pull what is needed. On the other side, too little current may or may not work, depending on the load. If the load has an internal battery, like a cellphone charger, it should work fine but charge a little slower. If the load is a heater or a light, it usually will work with just a little less heat or light being produced.

To size a simple DC system without a voltage regulator:

1. **Determine the DC-in voltage range** of the device you want to power. These are usually listed on the label of the device or the charger that conventionally powers the device. For example, the USB charger in Figure 6.1 shows a DC-in voltage range of 12-24 V, and the water pump in Figure 6.2 was experimentally shown to work quite well between 14 and 20 V. If the device needs exactly one voltage, then you will need a voltage regulator sized in Section 6.2 below.

2. **Determine the needed DC-in current** of the device you want to power. The max current needed is usually listed on the device. For example, the water pump in Figure 6.2. is rated at 7.5 A. Sometimes the device lists only the out current. For example, the USB charger in Figure 6.1. shows a DC-out current of two plugs with 5 V * 2.4 A, for a power of 12 W. If you want to power both plugs, that would be 2 plugs * 12 W/plug

for 24 W. If your minimum voltage is 12 V, then the maximum current needed (without accounting for efficiency loss) would be 2 A (because I=P/V for I=24 W/12 V). If the device lists only max power (in W), you can divide that power by the voltage (in V) to get max current (in A).

3. **Source a photovoltaic panel** whose power is sufficiently close to or more than the power needed by the device. If the device doesn't list the power it needs, multiply its listed max current by max voltage to know its max power needs.

   a. **The panel's max voltage** must be within the device's DC in voltage range from Step 1, or you will need a voltage regulator described in Section 6.2 below.

   b. **The panel's max current** should be close to or more the needed DC-in current of the device.

   c. Note that you may need to combine a few panels to meet your needs for a larger load. If so, see Section 7.3.1 for advice.

4. **Size the wire** as described in Section 7.6.

5. **You may want to add a switch** in order to be able to turn off the device when the sun is up and the panel is providing power. See Section 7.7 for sizing switches.

Specifications:

Rating:	12V-24V DC
Output:	2 Port USB 5V-2.4A/2.4A
Contact Resistance:	50m ohm Max.
Insulation Resistance:	DC 500V 100M ohm Min.
Operation Temperature:	-20°C ~ +45°C
Storage Temperature:	-20°C ~ +60°C

Figure 6.1

A 2 port USB that outputs 5 V at 2.4 A (so 12 W to each port) and accepts DC input between 12 and 24 V, which fits for many small PV panels (appropedia.org/Redwood_Coast_Montessori_solar_charging_station)

Figure 6.2

A Shurflo 4008-101-E65 12 V water pump rated at 7.5 A that was experimentally shown to operate quite well between 14 and 20 V from solar (appropedia.org/CCAT_solar_irrigation_for_food_forest).

6.2. Load with a voltage regulator

Often, powering a simple DC load without a battery will require a voltage regulator to bring the voltage of the panels into the right range for the device to be powered. This allows for all the simplicity and savings of a battery-less system without having to perfectly match the panel voltage. For devices that need a single specific voltage (i.e., not a voltage range), then a voltage regulator or a battery-based system with a charge controller will be needed. This book calls voltage regulators a subset of charge controllers. Simple voltage regulators are unique in that they do not require a battery, whereas all other charge controllers will generally need a battery.

To size a voltage regulator:

1. Source a regulator with a **DC-in voltage** range that fits your solar panel. For example, the voltage regulator in Figure 6.3 accepts 15-40 V, which fits this panel's output of 16-21 V.

2. Source a regulator with a **DC-out voltage** that matches the load's needed voltage. For example, the voltage regulator in Figure 6.3 outputs exactly 12 V, which is precisely what the LED requires.

3. Make sure the voltage regulator can provide the **needed current** of the load. For example, the LED in Figure 6.3 requires 0.36 Amps for full brightness. This voltage regulator can provide up to 2 A, which could power 5 LEDs (5 LEDs * 0.36 A/LED = 1.8 A).

Double check that your panel(s) can provide the needed current and do so in the required voltage range. You may need to put a few panels together to meet your needs. In this case, see Section 7.3.1 for advice. For example, the panel in Figure 6.3 is a 20 W panel with a maximum power point at about 18 V and 1.1 A. This may be able to fully power only 3 LEDs (3 LEDs * 0.36 A/LED = 1.08 A), depending on the efficiency and electronics of the voltage regulator. A high-quality voltage regulator might be able to power more based on power instead of current. In this case, the LEDs take 4.32 W (12 V * 0.036 A). This 20 W panel could power four LEDs based on power (4 LEDs * 4.32 W = 17.28 W).

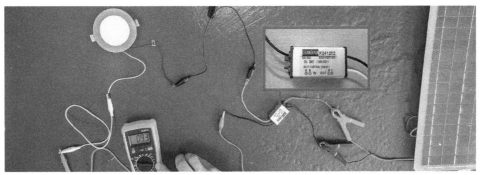

Figure 6.3
Solar panel powering an LED (left) through a voltage regulator (center) with no battery. This system could power more than one LED.

7. Sizing a system with batteries

A system with batteries has many advantages, such as access to energy even when the sun is not out. Sizing systems is usually an iterative approach, and if you are sizing a battery-based system, the steps are generally:

1. Needs assessment

2. Sizing batteries

3. Sizing panels

4. Sizing charge controller

5. Sizing inverter

6. Sizing wires and plugs

7. Sizing disconnects

Sizing a system often involves a lot of research, math, and understanding. You can definitely do this! We find it to be an enjoyable challenge that gets easier each time. You may want to check out the available sizing tools, calculators, and spreadsheets at https://www.appropedia.org/Solar_photovoltaic_software to augment or replace the work in this section.

7.1. Needs assessment (aka energy audit)

Any device that consumes electrical energy is called a load (e.g., a water pump, tv, lights, or computer). Include only the devices you want to power with solar or a battery in your design. You want to examine your own energy consumption and first reduce your energy needs as much as possible. This is because it will reduce the size of your PV system and reduce your initial costs. For example, change your lighting from incandescent to LEDs to greatly cut your energy use and thus the size of the PV system that you need. After you've found every means to reduce your energy use,[163] figure out how much power (in watts) each appliance or load uses and how much time you would normally plan to use it. Most appliances have the power they use on the back label. Other means of finding out how much power a device uses include looking at specification sheets found online, contacting local appliance dealers or the product manufacturers, general Google searches, or measuring the power use yourself with a power-meter device such as the Kill-a-Watt.[164] If you are sizing a system for a house and it is currently grid connected, your previous electricity bills will give a very good idea of how much energy your household consumes and will allow you to size a system which will offset some or all of the electricity consumption for your house.

The next step is to calculate your AC and DC loads. Sometimes you have more complicated systems that are half AC and half DC. It is easier to make an either all DC or all AC system. So, first, you want to list your AC loads, like your hair dryer, refrigerator, and computer. Then do the same for your DC loads, if you have any. Remember that loads that plug into the wall are AC loads. That said, some devices, e.g., laptop chargers, convert the AC to DC before plugging into the load.

DC loads can be added up on a weekly basis for an easy estimation. To do so, make a list of each DC load and its wattage. Then, multiply each load by the number of hours you use it per day and by the number of days you use it per week to calculate the Weekly Energy Use for each load. After that, add up all the individual DC Weekly Energy Uses

---

163 See more energy saving tips at https://www.appropedia.org/How_to_make_your_home_more_energy_efficient

164 Instructions for using a kill-a-watt at https://www.appropedia.org/How_to_use_a_KillAWatt_meter

for a total DC Weekly Energy Use, and divide by 7 days/week to calculate an estimate for your Total DC Daily Energy Use. In addition, add up the individual DC powers for a total Max DC Power, if all loads were on at the same time, or you can selectively add the maximum DC loads to be on at the same time. This Total DC Power value will be used for sizing wires later. Calculation 1 shows a blank DC Load Analysis.

Calculation 1
DC Load Analysis

DC Load	DC Power (W)	×	Daily Usage (hr/day)	×	Weekly Usage (days/week)	=	DC Weekly Energy Use (Wh/week)
		×		×		=	
		×		×		=	
		×		×		=	
		×		×		=	
		×		×		=	
		×		×		=	

A. Total DC Power				Total DC Weekly Energy Use		Wh / Week
				÷		7 Days / Week
				B. Total DC Daily Energy Use		Wh / Day

AC loads, similar to DC loads, can be added up on a weekly basis for easy estimation using the same steps described above.[165] However, since solar panels generate DC power, it must be converted through an inverter to meet the AC loads. Therefore, once you have calculated the Total AC Daily Energy Use, you will then divide this number by an inverter efficiency from the manufacturer[166] in order to calculate the Total DC daily demand from running the AC loads off of the inverter. In addition, add up the individual AC powers for a total Max AC Power, if all loads were on at the same time. Calculation 2 shows a blank AC Load Analysis.

---

165 For small systems with inverters that are always on, you may want to account for the power the inverter uses to be on. For many inverters, this value is just a few watts.

166 If you do not know the efficiency, an inverter efficiency of 0.9 is a safe value.

Calculation 2
AC Load Analysis

AC Load	AC Power (W)	×	Daily usage (hr/day)	×	Weekly usage (days/week)	=	AC weekly energy use (Wh/week)
		×		×		=	
		×		×		=	
		×		×		=	
		×		×		=	
		×		×		=	
		×		×		=	

C. Total AC Power		**Total AC weekly energy use**	**Wh/week**
		÷	7 days/week
		Total AC daily energy use	**Wh/day**
		÷ inverter efficiency	0.9
		D. Total DC daily inverter demand	**Wh/day**

Please note that some advanced appliances do not consume their rated power continuously and have various power cycles. For example, a refrigerator only consumes its full rated power for a couple of dispersed hours per day and uses much less at other times. These advanced appliances are better represented by entering a weekly energy value (measured using an energy meter or looked up e.g., from Energy Star), and a max power for their AC power (usually on the back label of the appliance as the rated power) instead of doing the math with the daily and weekly usage.[167]

Finally, add together your Total DC Daily Energy Use (from Calculation 1) and your Total DC Daily Inverter Demand (from Calculation 2) to get your Total Daily DC Demand, as shown in Calculation 3:

---

167   See much more at https://www.appropedia.org/How_to_do_an_electrical_energy_audit

Calculation 3
Total daily DC demand (Wh/day)

B. Total DC daily energy use	+	D. Total DC daily inverter demand	=	E. Total daily DC demand
	+		=	Wh/day

7.2.   Sizing Batteries

For an off-grid system, if you want power when the sun is not out, you will need storage (probably batteries) for your system. If you are planning on doing a grid-tied system, you can probably skip this step. However, if you need battery backup for a grid-tied place such as a hospital, you will also need to determine battery size.

If you are installing a system for a weekend occupied scientific outpost, you might want to consider a larger battery bank and a smaller PV array because your system will have all week to charge the battery and store the energy. Alternatively, if you are adding a solar panel array as a supplement to a generator-based system, your battery bank can be slightly undersized since the generator can be operated if needed for recharging.

To determine the total storage needed, you need to first determine how long you want the electricity to be provided if there is no sunlight. This is usually expressed as "days of autonomy" because it is based on the number of days you want your system to provide power without receiving an input charge from the solar panels or the grid. You also need to consider your usage pattern and the critical nature of your application. If it is not very important to you to have electricity every single day, then it is okay to use a small number, such as 1, for the days of autonomy.

Once you have determined your days of autonomy, multiply the total daily DC demand (Calculation 3, Calculation 4: Energy demand (Wh)-E) by the days of autonomy as shown in Calculation 4:

Calculation 4
Energy demand (Wh)

E. Total daily DC demand (Wh/day)	x	Days of autonomy (days)	=	F. Wh demand
	x		=	Wh

To size the batteries to meet this Wh, you will need to adjust the capacity of the battery based on temperatures (if it will get cold in the battery storage) and its maximum depth of discharge.

To adjust for temperate, select from Table 7.1 the closest multiplier for the average ambient winter temperature your batteries will experience.

Table 7.1
Battery capacity temperature multipliers[168]

Temp (°C)	Zinc-Chloride	Portable Sealed NiMH	Iron-Electrode	Lithium Ion	Lithium Iron Phosphate	Lead Acid
-20°C	-	-	-	0.51	-	0.64
-10°C	0.60	0.50	0.5	0.70	0.75	0.76
0°C	0.80	0.80	0.7	0.82	0.91	0.85
10°C	0.97	0.85	0.9	0.89	0.97	0.92
20°C	1.00	0.90	1.00	0.93	1.01	0.98
25°C	1	1	1	1	1	1
30°C	1.10	1.00	1.00	1.00	1.02	1.02
40°C	1.15	1.00	1.00	1.00	1.02	1.04

---

168   Find more information and further explanation of battery multipliers at https://www.appropedia.org/Battery_temperature_to_capacity_tables

Next, you will need the depth of discharge (DOD) limit for the batteries you are selecting. This DOD value is usually between 0.2 - 0.9 (i.e., between 20 and 90%). You can determine the DOD for your battery based on the manufacturer specifications or an average based on the type of battery, as shown in Table 5.2.

Once you have determined the temperature multiplier and depth of discharge, divide the calculated Wh demand by the depth of discharge limit, and multiply by the temperature multiplier to determine the needed battery storage size, as shown in Calculation 5:

Calculation 5
Needed battery storage (Wh)

F. Wh demand (Wh)	÷	Depth of Discharge	x	Temp mulitiplier	=	G. Needed battery storage
	÷		x		=	Wh

Often, batteries are rated in Amp hours (Ah). Remember that Power = Amps x Volts and the units are W=A*V. We can multiply both sides of this equation by hours to get Wh=Ah*V and divide both sides by V to get: Ah=Wh/V. So, to calculate required battery storage in Amp hours, simply divide the needed battery storage in Wh (Calculation 5-G) by the nominal battery bank voltage, as shown in Calculation 6.

Calculation 6
Needed battery storage (Ah)

G. Needed battery storage (Wh)	÷	H. Nominal battery bank voltage (V)	=	I. Ah needed
	÷		=	Ah

The nominal battery bank voltage will usually be 12 V, 24 V, or 48 V to match commonly available charge controllers. In addition, if you will be running your inverter or DC loads directly from the battery bank, the battery bank voltage must match the voltage of those loads (e.g., car inverters are 12 V, but many other inverters can also do 24 V and 48 V).[169]

---

169 Note that usually the nominal battery bank voltage is the same as the nominal photovoltaic array voltage, charge controller voltage, and inverter voltage. Most systems will have all parts at the same nominal voltage; that said, some charge controllers can handle a photovoltaic array at one voltage and a battery bank at a different voltage. Common voltages are 12 V, 24 V, and 48 V.

The larger the power needs, the higher the voltage you will want. Increasing the voltage for higher power needs allows the current to stay lower, which allows for smaller wires (resulting in lower costs). The same is true regarding the distance to your loads and inverter. It is safer and less expensive to raise the voltage (and thereby lower the current) for systems with loads and inverter farther from the battery.

The only steps remaining in sizing the battery storage revolve around the number of batteries in series and parallel based on the batteries you can purchase. You can start with the needed battery storage size in Ah and the system voltage to start searching for the right batteries, or you can start with the batteries you want to use. Often you will want to iterate a little bit.

For a small system, you can often size a single battery to match the system voltage and needed storage—which makes your math and wiring very easy with just one battery. If you find just one battery that meets your voltage and Ah (or Wh) needs, then you can skip sizing the battery bank and move on to sizing the panels.

7.2.1. Battery Bank

If you need multiple batteries to meet your storage needs, then you will need to design your battery bank (i.e., your set of batteries in series and/or parallel).

Batteries are arranged in series to add up to a system voltage (remember that voltage adds in series), and then those series are arranged in parallel to add up to needed storage (remember that current adds in parallel). For example, if you have 12 V batteries with 110 Ah of storage being combined to make a 48 V battery bank with 220 Ah of storage, you would arrange four batteries in series to add to the 48 V, and arrange two parallel legs to add up to the needed 220 Ah. Each parallel leg needs to have the same voltage, so this example would be eight total batteries: two legs in parallel and each leg with four batteries in series.

To determine the number of batteries in series, divide the Nominal Battery Bank Voltage (the rated battery bank voltage) by the nominal voltage of a single battery (i.e., the voltage one battery is rated at) as shown in Calculation 7:

Calculation 7
Batteries in series

H. Nominal battery bank voltage (V)	÷	Single Nominal Battery voltage (V/bat)	=	J. Batteries in series
	÷		=	batteries

Remember that this number should be an integer such as 1, 2, 3, etc., since battery voltage adds in series and you can't have a fraction of a battery in your system.

To determine the number of parallel legs, divide the needed battery storage in Ah by the single battery Ah as shown in Calculation 8:

Calculation 8
Parallel legs of batteries

I. Needed battery storage (Ah)	÷	Single battery Ah (Ah/leg)	=	K. Legs in parallel
	÷		=	legs

Again, this number needs to be an integer (such as 1, 2, 3, etc.) since battery current adds in parallel and you can't have a fraction of battery in your system. If this number would round down (e.g., 2.3 rounded down to 2), we suggest that you look for more ways to lower your energy use or a slightly larger battery before rounding up. Otherwise, you must round up in order to meet your demand.

Once you have your number of parallel legs, simply multiply the batteries in series by the number of legs in parallel to determine the total number of batteries, as shown in Calculation 9.

Calculation 9
Total batteries

J. Batteries in series (integer)	x	K. Legs in parallel (integer)	=	L. Total batteries
	x		=	batteries

Note that you will want to iterate with various sized batteries to find the battery bank size that best fits your needs and budget. Also keep in mind that this again should be an integer. Note, for safety and battery health do not mix-and-match batteries. In addition, it is safer to use larger batteries as opposed to more batteries in parallel due to wiring, exposure, and the chance of one faulted battery being discharged into by the other batteries.

7.3.   Sizing Panels

Panels provide the power for your system. To size panels for a system with batteries and/or grid-tie (Figure 7.1 and Figure 7.2), you will need to determine the panel size based on the Total Daily DC Demand.

Figure 7.1
A large house with a large system in Northern California.

Figure 7.2
A small grid tie, net zero home in Northern California.

System inefficiency increases your need for more panels. A typical conservative balance of system efficiency is 0.85 (i.e., 85%) which includes the losses brought in primarily from the charge controller and battery charging (wires and connections also play a role). The better the charge controller, the lower the system inefficiency. Using a MPPT charge controller will yield a higher system efficiency.

To determine the daily PV demand, you will divide the total daily DC demand (Calculation 3-E) by the balance of system efficiency as shown in Calculation 10:

Calculation 10
Daily PV demand

E. Total daily DC demand (Wh/day)	÷	Balance of System efficiency	=	M. Daily PV demand
	÷	0.85	=	Wh/day

If you have a battery, you may be storing, generating, and using energy at different times throughout the day. The panels receive their energy from the sun, so the more sun you have, the more energy you have. The value for the amount of sun you have is described in Section 3.1 and is based on a few factors and has a few names such as irradiation, insolation, or peak sun hours.[170] Determine the irradiation by researching your location and taking into account the angle of your panels.[171] The more normal (perpendicular) to the sun, the more irradiation you will receive. In addition, shadows on your panel will reduce your available sun. Shadowed areas should be avoided whenever possible, but when shadows are unavoidable, be sure to account for them when sizing your panels. You can use a tool like a solar pathfinder[172] to quickly plot the predicted shadows for each month at your location, or just make an estimate based on your knowledge of the sun path in the area. Values are often presented in units of hours/day of full sun or kWh/m²/day and can be found based upon your location and the angle at which you will mount the panels (Table 7.2).

Table 7.2

Example solar irradiation (insolation) data for a Arcata, California, USA, with its very cloudy winters and somewhat sunny summers in units of hours/day of full sun or kWh/m²/day.

Tilt (degree)	Jan	Feb	Mar	Apr	May	Jun	Jul	Aug	Sep	Oct	Nov	Dec	Avg
0	1.8	2.4	3.6	5.0	5.8	6.0	5.9	5.0	4.4	3.1	2.0	1.6	3.9
Latitude	3.0	3.4	4.4	5.3	5.5	5.4	5.4	5.0	5.1	4.1	3.2	2.8	4.4

To determine the panel sizing in terms of needed array power, you will divide the Daily PV demand (Calculation 10-M) by the Peak Sun Hours (this is the same value as irradiation aka insolation), as shown in Calculation 11.

You could run these calculations for each month or select your system sizing based on specific scenario. A common and useful scenario is to pick the month with the combination of the highest energy need and the lowest irradiation. For instance, if you have a flat panel system in the location referenced by Table 7.2 that powers lights you use every evening, it would be best to design around December with only 1.6 hours/day of full sun, since that is the month with the lowest daily sun. That way, you will always have sufficient power for your lights the rest of the year. To give another example, for

---

170   Note that often, irradiation is in the units of kWh/m²/day, but the value is the same for the units of hrs/day of peak sun.

171   Peak sun hours or insolation/irradiation tables can be found at https://www.appropedia.org/Climate_data#Solar_insolation. In addition, you will need to adjust for panel angle and any shading. Panel angle can be found in the solar irradiation data sources.

172   How to use a solar pathfinder at https://www.appropedia.org/Solar_pathfinder

designing a system with flat panels that will be used only in the summer (like a summer retreat center), you can design around May with 5.8 hours/day of full sun (Table 7.2).

Calculation 11
Needed array power

M. Daily PV demand (Wh/day)	÷	Peak sun hours (hr/day)	=	N. Needed Array Power
	÷		=	W

To determine the number of panels, simply divide the needed array power (Calculation 11-N) by the panel wattage you are considering purchasing or already have, as shown in Calculation 12.

Maybe you already know which panels you are using, maybe you have a discount, maybe you found some, or maybe you were waiting to see how much power you need before making your purchase. Now is your time to shine.

Calculation 12
Needed number of panels

N. Needed Array Power (W)	÷	Panel Wattage (W/panel)	=	O. Total Panels
	÷		=	panels

Total panels will need to be an integer as you cannot have a fraction of a panel. Try different panel wattages that are available to you in order to find the best fit for your system. Once done, round up to the nearest whole number. For a small system, you can often size a single photovoltaic to match the needed power—which makes your math and wiring very easy with just one panel. If you find just one panel that meets your power needs (W), then you can skip sizing the photovoltaic array (Section 7.3.1) and move on to filling out Photovoltaic array specification (Section 7.3.2).

7.3.1.   Photovoltaic array

If you need more than one panel (Calculation 12-O), then you will need to arrange your photovoltaic array in series and/or parallel. If you sized your system such that one panel is the correct wattage, then you can skip this section.

Remember that voltage adds in series and current adds in parallel. If you want multiple parallel legs, then each needs to have the same number of panels connected in series. Therefore, the total number of panels equals the number of panels connected in series times the number of parallel legs, as shown in Calculation 13.

For example, an array of six panels could be arranged into:

✳ one leg of six panels in series (highest voltage, lowest current)

✳ two parallel legs of three panels in series

✳ three parallel legs of two panels in series

✳ or six parallel legs each with just one panel (lowest voltage, highest current)

The same total power will be produced by any of these arrangements.

Calculation 13
Photovoltaic array series and parallel panels

P. Panels in series	x	Q. Parallel legs	=	O. Total Panels	
	x		=		panels

The product of panels in series times parallel legs (Calculation 13-O) should equal the total panels from Calculation 12-O.

You will need to iterate the arrangement of panels based on available charge controllers. You will need to make sure that the total open circuit voltage of the panels in series is less than the charge controller's maximum photovoltaic input voltage, and that the total current of the legs in parallel (times a safety factor, typically 1.25) is under the charge controller's maximum photovoltaic input current. In addition, panels in series will require smaller wires and fuses than panels in parallel since wire and fuse size is based on current.

Once you determine the arrangement, which is often an iterative approach, you can fill out the photovoltaic array specifications that will inform many of the remaining parts of your system.

7.3.2. Photovoltaic array specification

Determining your array specifications requires knowing your number of panels in series and legs in parallel (which might be just one, if you sized a small system) and multiplying that by the manufacturer specifications (value shown in Figure 4.8).

To determine the array open circuit voltage, multiply the number of panels in series by the manufacturer specified open circuit voltage (V_{OC}) of your panel(s):

Calculation 14
Array open circuit voltage (V_{OC})

P. Panels in series	x	Panel V_{OC} (V)	=	R. Array V_{OC}	
	x		=		V

To determine the array short circuit current, multiply the number of parallel legs by the short circuit current (I_{SC}) and a safety factor (typically 1.25). This safety factor is to account for situations where the panels are overproducing. These situations might be from reflected sun of a nearby building or lens effect from passing clouds that concentrate the irradiance from the sun to be more than 1000 W/m^2.

Calculation 15
Array short circuit current with safety factor (I_{SC})

Q. Parallel legs	x	Panel I_{SC} (A)	x	Safety factor	=	S. Array safety current	
	x		x	1.25	=		A

These values will come into play in the following sections.

This might be a good time to refresh on voltage adding in series and current adding in parallel from Section 4.2.

7.4. Sizing Charge Controller

Charge controllers regulate the voltage from your panels to prevent destruction of other system components such as batteries, loads, and inverters. In addition, some charge controllers can also protect the batteries from being drained too low by the loads. Most systems will want the charge controller to regulate the array current and voltage (Section 7.4.1) to protect the system. Many systems will also want the charge controller to regulate the load currents to protect the batteries from discharging too low (Section 7.4.2). In addition, make sure your charge controller is rated to work with whatever battery type you choose, as different battery types require different charging profiles (e.g., flooded lead acid vs. Lithium Iron Phosphate batteries).

7.4.1. Charge Controller Array Current and Voltage

The first step is to size the charge controller for maximum array current[173] and voltage. You have already determined these data in Calculation 15-S and Calculation 14-R.

---

173 If you are selecting an MPPT charge controller, you do not need to include the safety factor. Therefore, the array current can just be the number of legs times the short circuit current in Calculation 15. This may result in some lost energy (for current above the MPPT current), but will not break the system.

Table 7.3
Controller PV array current and voltage

		S. Array safety current	
Controller PV Array Current	=		A
		R. Array V_{OC}	
Controller PV Array Voltage	=		V

You will also want to confirm that your charge controller matches the battery type and nominal battery bank voltage from Calculation 6-H. If you are using only the charge controller to regulate the power from the panels, then you are done sizing the controller and can skip to Section 7.5-Sizing Inverter.

7.4.2. Charge Controller Load Current

If you want the charge controller to also manage the loads, then you will need to size the loads for the controller. Higher power loads will often pull more current than the charge controller can handle. Those higher power (and therefore higher current) loads can be connected in parallel with the battery instead. Those loads connected in parallel with the battery will no longer be controlled by the charge controller but will still have the regulated voltage and stored energy of the battery.

You can pick which devices you want to allow the charge controller to control and connect only those devices to the charge controller. Connect the rest of the devices directly to the fused battery. The controller load sizing may be too large if you base this on the maximum power from the load calculation sheets.

It is possible that your system will never have all devices running at maximum power all at the same time, in which case add up the max AC and max DC power based upon your knowledge of the maximum devices running simultaneously.

To calculate charge controller max load current from the inverter, divide the AC power to control (Calculation 2-C)[174] by the Inverter voltage (which is usually 12 V, 24 V, or 48

---

174 Remember that this can be all of the AC loads on at once from the needs assessment, or some smaller number of loads.

V and should match the battery voltage) and the inverter efficiency used in Calculation 2[175]:

Calculation 16
Controller - DC current load from inverter

C. Total AC Power (W)	÷	Inverter voltage (V)	÷	Inverter efficiency	=	T. Max DC Load for AC
	÷		÷	0.9	=	A

To calculate the Charge controller max load current from purely DC devices, divide the total DC power to control[176] (Calculation 1-A) by the DC voltage (usually 12 V, 24 V, or 48 V and should match the battery voltage):

Calculation 17
Controller - DC current load from DC loads

A. Total DC Power (W)	÷	DC Voltage from controller (V)	=	U. Max DC Load for DC
	÷		=	A

Finally, the charge controller max DC load is just the sum of the max load from the inverter for the AC loads and the max load from the purely DC loads:

Calculation 18
Total controller DC current load

T. Max DC Load for AC	+	U. Max DC Load for DC	=	V. Total Max DC Load (A)
	+		=	A

It is very likely that this total max DC load will be too large for your charge controller. If it is too large, you can size a larger controller, connect multiple charge controllers in parallel (if the system architecture allows), find more efficient loads, engage more conservative use patterns, or, more likely, connect some or all of the loads in parallel with the battery instead of controlling from the charge controller. Remember that if

---

175 Pearsall, N. (Ed.). (2016). The performance of photovoltaic (PV) systems: Modelling, measurement and assessment. Cambridge, England: Woodhead Publishing. Figure 1.4 https://doi.org/10.1016/C2014-0-02701-3

176 Remember that this can be all of the DC loads on at once from the needs assessment, or some smaller number of loads.

you do connect your loads directly to the battery, you will not have anything protecting your battery from over-discharging. Over-discharging your batteries could significantly reduce their lifetime, so be sure to manually monitor your batteries' voltage and adjust your power consumption to maintain a proper voltage (typically above 11 V for a 12 V battery bank). Note that unlike most loads, most inverters do have a low voltage disconnect feature built in.

7.5. Sizing Inverter

An inverter converts the DC power produced from the solar panels into AC power for your typical plug-in devices. There are a few factors to consider when sizing your inverter. For example, you only need to power the maximum wattage that will be running simultaneously. For some systems, this will be lower than the total AC power (e.g., a system that has only lights on at night and only fans on during the day). In addition, some devices have a large surge power requirement (e.g., power tools). Inverters are also rated on the total surge current they can handle.

Sizing an inverter is usually as simple as multiplying the max AC power (Calculation 2-C)[177] by an inverter safety factor of 1.25:

Calculation 19
Inverter size in watts

C. Total AC Power (W)	x	Inverter Safety Factor	=	W. Inverter Size in Watts
	x	1.25	=	W

---

177 You can also choose to allow the powering up of only a fraction of the AC loads at once in order to design a smaller system...but you will need to make certain that more loads are not turned on at once.

If you want the inverter sized in Amps, just divide by the inverter voltage, which is usually the same as nominal battery bank voltage (and always is the same if the inverter is connected to the batteries instead of the charge controller):

Calculation 20
Inverter size in Amps

W. Inverter Size in Watts	÷	Inverter voltage (V)	=	X. Inverter Size in Amps
	÷		=	A

You will want an inverter that matches the inverter voltage and meets or exceeds the inverter size requirement.

Note: If you have large surge devices, such as power tools, you will want to also add up all the surge watts and make sure your inverter is rated to cover those.

7.6. Sizing Wires and Plugs

Wires and plugs convey the power to where you want it. Wires and plugs are rated for their voltage and current. Plugs can also be specific to the voltage and AC versus DC.

Wires are for electrical charge like pipes are for water. For both wires and pipes, they can only handle so much pressure (voltage in electricity), and their diameter (gauge in electricity) determines how much current can pass through. The math for this relationship of wire material, diameter, and length is interesting and is quite similar for pipes, but we suggest using a traditional wire length table such as Table 7.4 (also called an ampacity chart). In order to keep your costs low, you will want to keep your wire lengths short and your currents low.

Wire length tables are based on voltage, current, and length. You will look at the voltage and maximum current for each place you need wire, e.g., from the solar panels to the charge controller, from the charge controller to the batteries, from the charge controller to the loads, etc. These values can be found in the previous sizing sections.

For example, from Table 7.4, if the maximum current running your inverter is 15 A, and the inverter is 10 feet away, then you will need 10-gauge wire because the roundtrip length is 20 feet.

Table 7.4

Wire size based on round trip length for various currents in a 12 V DC system.[178]

RT length	AWG size based on Max Current (A) for a 12 V DC System (3% max loss)							
	5 A	10 A	15 A	20 A	25 A	30 A	40 A	50 A
15 ft	16	12	10	10	8	8	6	6
20 ft	14	12	10	8	8	6	6	4
25 ft	14	10	8	8	6	6	4	4
30 ft	12	10	8	6	6	4	4	2
40 ft	12	8	6	6	4	4	2	2
50 ft	10	8	6	4	4	2	2	1
60 ft	10	6	4	4	2	2	1	1/0
70 ft	8	6	4	2	2	2	1/0	2/0
80 ft	8	6	4	2	2	1	1/0	2/0
90 ft	8	4	2	2	1	1/0	2/0	3/0

You will want to use plugs that are standard for your current and voltage so that they do not get accidentally misused. Similar to wires, you will assure that the plug you are using is rated to meet or exceed the voltage and current that will be connected to the plug. Dissimilar to wires, you will not need to consider the lengths.

---

178 Find more tables and formula at https://www.appropedia.org/Wire_length_tables

7.7. Sizing Disconnects

Disconnects act as a switch to turn on (close the circuit) or turn off (open the circuit) your current. Disconnects can be placed for safety, e.g., disconnecting the batteries or panels, or for convenience, e.g., turning your loads on and off. To determine the size of your disconnects, you will use the voltage and maximum current for the location of each disconnect.

Below are the most commonly disconnected items and how to size them:

For a **PV array disconnect,** you will pick a disconnect that can handle the array max current and max voltage from Section 7.3.2-Photovoltaic array specification. The current is usually the limiting factor and was determined with a safety factor in Calculation 15-S. A PV array disconnect is important for safety and system maintenance.

For a **battery disconnect,** you will pick a disconnect that is equal to or larger than the charge controller current. If you have other loads connected to your battery, make sure that they are under the charge controller current, or connect them to the battery with a separate fuse. A battery disconnect such as a fuse or breaker helps protect from shorting of the battery…and the battery is the most dangerous component in your system.

For a separate **inverter disconnect,** you will pick a disconnect that can handle the inverter Amps found in Calculation 20-X. An inverter disconnect is usually unnecessary, as many inverters contain their own fuse.

If you would like a fuse on an **individual load,** use the voltage and calculate the maximum current for that load (max power in watts divided by voltage) to determine the disconnect size. You may desire a disconnect for an individual load for convenience (e.g., a light bulb) or safety (e.g., a plug with a large load or water danger, in which case you would want a ground fault control interrupt).

8. Other useful stuff

8.1. Using old panels

An increasing number of old panels are entering the waste-stream and the marketplace. This will most likely continue as more systems get installed and efficiencies and costs keep improving. We have examined many panels that after twenty years are still producing over 80% of their specified maximum power. Here are some notes on using old panels:

1. **Quick check**: Check the I_{SC} and V_{OC} using a multimeter with the panel perpendicular to full sun. These are easy to check and give a sense of whether the panel is worth examining more. Make sure that your multimeter is rated for a higher current than the I_{SC} listed on the panel.

2. **More in-depth check:** To check the full health of the panel, you can construct an IV curve (Section 4.4) using a rheostat, or you can find a load (like an incandescent light) whose needed current is near the I_{MPP} listed on the panel. That said, if you know exactly what you want it to run, you can just test to make sure it can run the load and call it good.

3. **Combining panels:** If you want to combine panels, go back to Section 4.2 on series and parallel. You want each of your legs to have a similar voltage and each of your panels in a leg to have a similar current.

8.2. Units related to electricity

As you probably have noticed, the world is filled with important units and prefixes related to electricity and photovoltaics. The following tables are non-exhaustive lists of the most relevant units.

Table 8 .1
Electricity units[179]

Parameter (symbol)	Measuring Unit (symbol)	Electrical Description	Water Analog
Voltage (V)	volt (V)	Pressure (Potential) difference due to charge difference. $V=I*R$	Head: Pressure (Potential) difference due to height difference
Current (I)	amp (A)	Flow of charge in charge/time or coulombs/sec. $I=V/R$	Flow: Flow of water in volume per time such as liters/sec
Resistance (R)	ohm (Ω)	Opposition to the flow of charge. $R=V/I$	Friction: Opposition to the flow of water
Power (P)	watt (W)	Rate or work or energy transfer, storage, etc. Power=Energy/Time also $P=I*V$	Power: Power=Flow (Q) * Pressure (H)
Energy (E)	watt-hour (Wh)	The ability to do work. $E=P*t$ (where t is time)	Energy: The ability to do work

---

179 See https://www.appropedia.org/Plug_type_and_voltage_by_country the plug type, voltage, and frequency used in the region you are designing for.

Table 8.2
Multiples and sub-multiples of electricity units[180]

Prefix	Symbol	Multiplier	Power of Ten	Power symbol	Description (order of magnitude)
Tera	T	1,000,000,000,000	10^{12}	TW	Order of magnitude needed to power world
Giga	G	1,000,000,000	10^9	GW	The size of PV manufacturing facilities in China
Mega	M	1,000,000	10^6	MW	The size of a large solar farm is around 10 MW
Kilo	k	1,000	10^3	kW	The size of systems that can completely power a home
None	none	1	10^0	W	Enough power for a handheld device
Centi	c	1/100	10^{-2}	cW	A very low power device – e.g., LED flashlight
Milli	m	1/1,000	10^{-3}	mW	A typical hearing aid uses less than 1 mW
Micro	μ	1/1,000,000	10^{-6}	μW	Very small solar cells used as sensors, a few dozen needed for medical implants

Here are a few more units used in relation to photovoltaics:

AM1.5 spectrum - The American Society for Testing and Materials (ASTM) G-173 spectra represent terrestrial solar spectral irradiance on a surface of specified orientation under one and only one set of specified atmospheric conditions. The conditions selected were a reasonable average for the 48 contiguous U.S. states over a period of one year. The tilt angle selected is approximately the average latitude for the contiguous U.S.A.[181]

kW(p) - the peak power of a PV system when it is fully perpendicular to the sun on a cloudless day (e.g., under STC).

Standard Test Conditions (STC) - room temperature, 1000 W/m² of solar energy with an AM1.5 spectrum.

---

180 See https://www.appropedia.org/Powers_of_10 for more.

181 https://www.nrel.gov/grid/solar-resource/spectra-am1.5.html

Kilowatt-Hour (kWh) - the amount of electrical energy consumed by an electric circuit over a period of time. It is equal to 1000 Wh and generally what is on home electric bills.

8.3. Economics

In the end, one of the primary reasons people adopt solar PV for their applications is because of economics. PV economics can be challenging to understand because the technology lasts so long that discount rates and interest rates can play a major role, and most people simply do not understand them. At the end of the day, you want to know: will solar make electricity less costly than other alternatives? The good news is that for the vast majority of the world and the majority of applications, the answer is "Yes!" For some applications, it is a big "YES!" – where solar completely trounces the cost of existing sources of electricity. For example, any application using conventional retail purchased batteries is going to be run for much less money with a small solar PV system.

8.3.1. Cost of disposable battery per kWh

A Duracell AA battery[182] (alkaline-manganese dioxide battery MN1500) at a constant power draw of 250 mW will last about 10.5 hours, which is about 2,625 mWh or 2.625 Wh or only 0.002625 kWh. If you use the batteries more intensively, then you get even less electricity. For example, at 500 mW, the battery will last less than four hours, which is less than 2000 mWh. The best deal we could find on Amazon[183] was a 48-battery pack for US$18.99 or US$0.40 per battery. So, US$0.40/0.002625 kWh = US$152.38/kWh!! It is economic insanity to use disposable batteries for any application. Solar by comparison is only a few cents per kWh. How do we know that when we buy a solar cell, not the solar electricity? We need to calculate the levelized cost of electricity.

---

182 https://www.duracell.com/en-us/techlibrary/product-technical-data-sheets

183 https://www.amazon.com/Energizer-Batteries-Double-Battery-Alkaline/dp/B079GXSFPB/

8.3.2. Levelized cost of electricity

For bigger systems, the economic feasibility of PV projects is increasingly being evaluated using the levelized cost of electricity (LCOE) generation in order to be compared to other electricity generation technologies. Unfortunately, there is a lack of clarity of reporting assumptions, justifications, and degrees of completeness in LCOE calculations, which produces widely varying and contradictory results. A recent review[184] of the methodology of properly calculating the LCOE for solar PV has shown that PV is economical almost everywhere and will be used for the basis for our discussion here.

Given the state of the art in PV technology and favorable financing terms, it is clear that PV has already obtained grid parity in most locations. As installed costs continue to decline, grid electricity prices continue to escalate, and PV industry deployment experience increases, PV will become the dominant source of electricity over expanding geographical regions on the main grid. But what about your own project? How do you calculate LCOE for it?

The LCOE requires considering the cost of the energy generating system and the energy generated over its lifetime to provide a cost in $/kWh (or $/MWh or cents/kWh). We will introduce three methods: 1) back of the envelope, 2) LCOE spreadsheet, and 3) The System Advisor Model (SAM). These are in order of complexity and reliability. So, for example, if you just want to know if you should invest in a single mini-module to charge your cellphone, method 1 is probably more than good enough. If you are looking at a system for yourself, method 2 will suffice. Finally, if you are planning to roll out a major bank-funded series of PV systems across many communities, method 3 (SAM) would be the best choice.

8.3.3. Back of the Envelope Method

To get a basic idea on the LCOE of your PV system, you can do the math on the back of an envelope.

---

184 K. Branker, M.J.M. Pathak, J.M. Pearce, A Review of Solar Photovoltaic Levelized Cost of Electricity, *Renewable and Sustainable Energy Reviews*, *15*, pp.4470-4482 (2011). https://mtu.academia.edu/JoshuaPearce/Papers/1540664/A_Review_of_Solar_Photovoltaic_Levelized_Cost_of_Electricity

The basic equation for an LCOE = Total Cost / Total Energy Generated.

Total Cost can be assumed to be the capital cost of your PV system. The Total Energy Generated = Lifetime of the system in years × the energy generated from your system per year.

Let us assume that you bought a PV system consisting of a 200 W module and a small battery for $350 and that you live in the middle of India. If we look at Figure 3.23, it looks like you get about 1700 kWh/m²/year or that many sun hours in a year. Most PV will easily last for 20 years or more. So, the total energy you would expect to generate would be 20 years x 1700 kWh/kW/year x 0.2 kW=6,800 kWh

The cost per kWh is thus $350/6,800 kWh = $0.05/kWh

This is a rough idea and is the simplest case where you have the cash (e.g., no financing necessary), and no degradation, etc.

If you would like to include those variables, then you should try the Spread Sheet Method.

8.3.4. Spreadsheet Method

Recognizing that LCOE is a benchmarking tool, there is high sensitivity to the assumptions made, especially when extrapolated several years into the future. Thus, assumptions should be made as accurately as possible. Especially in the case of renewable energy technologies, like solar PV, that are capital intensive with negligible maintenance (e.g., there are no fuel costs), it is important to make the appropriate assumptions when comparing systems.

When reporting an LCOE for solar PV for yourself or others, you should be clear in the inclusion of assumptions and specifications which make each calculation unique. Thus, when a value is reported, it should also clearly include:

1. The solar PV technology and degradation rate (e.g. c-Si or a-Si:H, and 0.5%/year degradation rate, etc.).

2. Scale, size, and cost of PV project [including cost breakdown] (residential, commercial, utility scale/# kW, # MW, $/Wp).

3. Indication of solar resource: capacity factor (a measure of the solar energy collected in terms of energy generated divided by the installed capacity), solar insulation, geographic location, and shading losses.

4. Lifetime of the project and terms of financing (these are not necessarily equal).

5. Financial terms: financing (interest rate, term, equity/debt ratio, cost of capital), and discount rate.

6. Additional terms: inflation, incentives, credits, taxes, depreciation, carbon credits, etc. (these need not be in the analysis, but it should be stated whether or not these are included).

Thus, the authors would suggest the degree of applicability of their analysis so that sweeping assumptions as to their decisions are not incorrectly made.

On Appropedia, a free simple open source calculator is provided for finding the LCOE of a solar PV system.[185] The default scenario is for a system in Kingston, Ontario, Canada. The "assumptions and sources" section in the calculator gives guidelines on how to change inputs based on location. The primary equations are shown in Figure 8.2. Readers without a detailed mathematics background should not be intimidated by the equation. We purposefully embedded it in a spreadsheet that is carefully locked to avoid anyone messing up the primary calculation.

---

185 https://www.appropedia.org/File:ECM032_-_Solar_PV_LCOE.xls

$$LCOE = \frac{\sum_{t=0}^{T} \dfrac{C_t}{(1+r)^t}}{\sum_{t=0}^{T} \dfrac{E_t}{(1+r)^t}}$$

Nomenclature

T	life of the project [years]
t	Year t
C_t	Net cost of project for t [$]
E_t	Energy produced for t [$]
I_t	Initial investment/ cost of the system including construction, installation etc. [$]
M_t	Maintenance costs for t [$]
O_t	Operation costs for t [$]
F_t	Interest expenditures for t [$]
r	Discount rate for t [%]
S_t	Yearly rated energy output for t [kWh/yr]
d	Degradation rate [%]

$$LCOE = \frac{\sum_{t=0}^{T} \dfrac{I_t + O_t + M_t + F_t}{(1+r)^t}}{\sum_{t=0}^{T} \dfrac{E_t}{(1+r)^t}} = \frac{\sum_{t=0}^{T} \dfrac{I_t + O_t + M_t + F_t}{(1+r)^t}}{\sum_{t=0}^{T} \dfrac{S_t(1-d)^t}{(1+r)^t}}$$

| 1.Home | 2.Introduction | 3.Inputs | 4.Executive Summary | 5.Projected Savings | **6. Assumptions & Sources** |

Figure 8.2
LCOE spread sheet screen shot showing tabs and primary equation.

This spreadsheet consists of six sheets, or tabs, including this Home sheet. The Intro sheet provides an overview of the LCOE and an example. On the Input sheet, enter the best estimates for your project in the orange cells (and change default purple cells if required) (See Table 8.3 for color coded legend). A summary of your costs and benefits based on the values entered in the Input sheet are provided on the Executive Summary sheet. The detailed economic and environmental analyses behind the executive summary results appear on the Savings Projection sheet, which shows financial and environmental metrics. Assumptions underlying the calculations and explanations of the defaults (such as electricity costs and emissions) are found on the Assumptions & Sources sheet (Figure 8.2) and may need to be adjusted for your application. The LCOE calculator has been designed primarily for electronic use but can also be printed if necessary. Printing defaults have been set up such that all pages will fit on standard Letter or A4 paper and will rotate (landscape vs. portrait) automatically as necessary.

Table 8.3
Legend for LCOE spreadsheet calculator.

Legend		
	Headings	green
	Input Cells	orange
	Default Values	purple
	Summary Charts	blue

8.3.5.   SAM Method

The System Advisor Model[186] (SAM) is designed for people involved in the renewable energy industry to make good financial decisions. It was developed by the National Renewable Energy Laboratory (NREL) in the U.S. and has been made open source.[187] NREL will continue to maintain and update the code and to release NREL versions of SAM for Windows, Mac, and Linux, with new features and updates. You can download it for free.[188]

SAM makes performance predictions and cost of energy estimates for renewable power projects based on installation and operating costs and system design parameters that you specify as inputs to the model. The first step in creating a SAM file is to choose a technology and financing option for your project. SAM automatically populates input variables with a set of default values for the type of project. Just as with the Spreadsheet Method, it is your responsibility to review and modify all of the input data as appropriate for your project. Next, you provide information about a project's location, the type of equipment in the system, the cost of installing and operating the system, and financial and incentive assumptions.

To describe the renewable energy resource and weather conditions at a project location, SAM requires a weather data file. Depending on the kind of system you are modeling,

---

186 https://sam.nrel.gov/

187 https://sam.nrel.gov/opensource

188 https://sam.nrel.gov/download

you either choose a weather data file from a list, download one from the Internet, or create the file using your data.

SAM includes several libraries of performance data and coefficients that describe the characteristics of system components such as photovoltaic modules and inverters, wind turbines, and biopower combustion systems. For those components, you simply choose an option from a list, and SAM applies values from the library to the input variables.

For the remaining input variables, you either use the default value or change the input variables:

✳ Installation costs including equipment purchases, labor, engineering and other project costs, land costs, and operation and maintenance costs

✳ Numbers of modules and inverters, tracking type, derating factors for photovoltaic systems.

✳ Collector and receiver type, solar multiple, and storage capacity

✳ Analysis period, real discount rate, inflation rate, tax rates, internal rate of return target, or power purchase price for utility financing models

✳ Building load and time-of-use retail rates for commercial and residential financing models

✳ Tax and cash incentive amounts and rates

Once you are satisfied with the input variable values, you can run simulations and then examine the results. A typical analysis involves running simulations, examining results, revising inputs, and repeating that process until you understand and have confidence in the final results.

SAM displays simulation results in tables and graphs, ranging from the metrics table displaying first year annual production, and other single-value metrics, to the detailed annual cash flow and hourly performance data that can be viewed in tabular or graphical form.

A built-in graphing tool displays a set of default graphs and allows for creation of custom graphs. All graphs and tables can be exported in various formats for inclusion in reports and presentations and also for further analysis with spreadsheets or other software.

SAM's performance models make hour-by-hour calculations of a power system's electrical output, generating a set of 8,760 hourly values that represent the system's electricity production over a single year. SAM includes performance models for PV systems (flat-plate and concentrating), battery storage for PV, solar water heating, wind power (large and small), geothermal power and geothermal co-production, and biomass power, among other renewable energy technologies.

SAM's financial model calculates financial metrics for various kinds of power projects based on a project's cash flows over an analysis period that you specify. The financial model uses the system's electrical output calculated by the performance model to calculate the series of annual cash flows. It also includes a simple levelized cost of energy calculator based on a fixed charge rate input.

Learning how to use SAM is for those that are planning a lot of system designs. If you are just doing a small system for yourself, it is probably far more granular than you need. If you want to make the investment, however, the learning is free. To learn how to use SAM, there are free webinars, sample files, and curriculum materials.[189]

8.4. Disclaimer

The publisher and authors provide the information in this book, and its references, for educational purposes. We advise you to take full responsibility for your safety. Before implementing the skills described in this book, be sure to follow safe practices listed in

---

189 https://sam.nrel.gov/resources

this book and from your other sources and knowledge regarding working with electricity. Do not take risks beyond your comfort level.

You should understand that when working with electricity, there is the possibility of physical injury and death. If you engage in photovoltaic work, you agree that you do so at your own risk, assume all risk of injury to yourself, and agree to release and discharge the publisher and the authors from any and all claims or causes of action, known or unknown, arising out of the contents of this book.

If you are just getting started, we strongly suggest you start small and build your skills while incurring less risk. **Please be safe, have fun, and positively impact your world!**

8.5. Problem sets

A list of problems to develop your knowledge or to integrate into a class are available at https://www.appropedia.org/To_Catch_the_Sun#Problemsets. Teachers, email lonny@humboldt.edu for an answer guide.

8.6. Example systems and online community

We suggest you find local and global communities to work with. Mutual aid organizations are a great place to get involved.

We rely heavily on Appropedia for iterating designs and you can explore example working PV systems at https://www.appropedia.org/To_Catch_the_Sun#Examples. Some great groups exist online and depending on the time you are reading this and the language you are reading it in, we suggest doing a search. In addition, a group we have loved working (and that significantly helped fundraise for this book) is Permies. com. You can ask all types of photovoltaic, and permaculture, related questions at https://permies.com/forums

9. Index

Total DC Daily Inverter Demand, 152–53
Total DC Power, 166
Total Energy Generated, 176
Total Max DC Load, 166

U
UNIBE (Universidad Iberoamericana), 13–14
units, 68, 85, 92, 141, 155, 160, 172–73
Universidad Iberoamericana (UNIBE), 13–14
Universidad Tecnológica de Coahuila (UTC), 27
USB charger, 100, 137, 143–44

V
VMPP, 94
voltage and current, 92, 144, 168–69
voltage regulator, 99–101, 104, 114, 136–37, 144–47
volts, 78, 105, 120, 155, 172

W
water conservation, 47
Waterpod, 186
water pump, 54, 132, 136–38, 144, 146, 150
watt-hours, 105
waveforms, 114–17
Wh, 82, 105, 154–56, 172, 174
wires and plugs, 89–90, 118, 168
wooden solar panel mount, 127
Wp, 49, 72, 80, 82, 94

10. Author bios

10.1. Joshua Pearce

Professor Joshua M. Pearce received degrees in chemistry, physics, and a Ph.D. in materials engineering from the Pennsylvania State University for his work in low-cost solar cells. He then developed the first Sustainability program in the Pennsylvania State System of Higher Education and helped develop the Applied Sustainability graduate engineering program while at Queen's University in Canada.

He was the first Richard Witte Professor cross-appointed in the Department of Materials Science and Engineering and the Department of Electrical & Computer Engineering at Michigan Technological University where he ran the Open Sustainability Technology Research Group. He was a Fulbright-Aalto University Distinguished Chair and remains visiting professor of Photovoltaics and Nanoengineering at Aalto University, Finland and Visiting Professor Équipe de Recherche sur les Processus Innovatifs (ERPI), Université de Lorraine, France. He wrote this book deep in the north in a solar-powered house.

He is currently the John M. Thompson Chair in Information Technology and Innovation at the Richard Ivey School of Business and the Department of Electrical and Computer Engineering at Western University in Canada.

Dr. Pearce is a frequent contributor to Appropedia, as his research concentrates on the use of open source appropriate technology to find collaborative solutions to problems in sustainability and poverty reduction. His research spans areas of electronic device physics and materials engineering of solar photovoltaic cells as well as solar photovoltaic systems and economics. His group also does work in open source RepRap 3-D printing, applied sustainability, and energy policy. His research is regularly covered by the international and national press, and it is continually ranked in the top 0.1% on Academia.edu. He is the author of the *Open-Source Lab: How to Build Your Own Hardware and Reduce Research Costs* as well as the *Create, Share, and Save Money Using Open-Source Projects.*

10.2.   Lonny Grafman

Lonny Grafman is an Instructor of Environmental Resources Engineering and Appropriate Technology at Humboldt State University; the founder of the Practivistas resilient community technology program; the Advisor and Project Manager (and at times fundraiser) for the epi-apocalyptic city art projects Waterpod, Flock House, WetLand, and Swale; the managing director of the north coast hub of BlueTechValley, supporting energy saving entrepreneurs; the director of the AWEsome Business Competition for groups working on Agriculture, Water, and Energy in Northern California; and the Founder and President of the Appropedia Foundation, sharing knowledge to build rich, sustainable lives.

Lonny has taught university courses and workshops in dozens of countries. He has worked, and led teams, on hundreds of domestic and international projects across a broad spectrum of sustainability - from solar power to improved cookstoves, from micro-hydro power to rainwater catchment, from earthen construction to plastic bottle schoolrooms. Throughout all of these technology implementations, he has found the most vital component to be community.

His first books, *To Catch the Rain* and *Atrapando la lluvia,* cover inspiring stories of communities coming together to catch their own rain, and how you can do it too.

Thank you for being part of this.
We appreciate you.
We would love you to be even more involved.
To that end, please:

✓ Sign up at **tocatchthesun.com** for the most up-to-date information and to learn about upcoming books.

✓ Spread the word to people you know, on social media (with the hashtag **#tocatchthesun**), and by writing a review on your favorite book/retail site.

✓ Share your projects and photos with us on social media or by adding them directly to **appropedia.org/To_Catch_the_Sun#projects**

✓ Follow us at **appropedia.org/To_Catch_the_Sun#About_the_Authors**

✓ Consider donating to Appropedia for To Catch the Sun related projects, or to any organization working on Solar power and let them know about the book! **appropedia.org/Appropedia:Support**

Joshua Pearce

Lonny Grafman